U0248397

SHI NOTES ON ARCHITECTURE

建筑拾记

Ten Years · Gathered · Ten Notes

朱 颖 / 王 鹏 / 张达志 等 编著
Zhu Ying / Wang Peng / Zhang Dazhi

图书在版编目（CIP）数据

建筑拾记 / 朱颖等编著. -- 天津：天津大学出版社，2018.1
ISBN 978-7-5618-6056-4

Ⅰ. ①建… Ⅱ. ①朱… Ⅲ. ①建筑设计 -- 中国 -- 文集 Ⅳ. ① TU2-53
中国版本图书馆 CIP 数据核字（2018）第 005498 号

Jianzhu Shi Ji

策划编辑：金　磊　韩振平
责任编辑：郭　颖
版式设计：朱有恒

出版发行　天津大学出版社
地　　址　天津市卫津路92号天津大学内（邮编：300072）
电　　话　发行部：022-27403647
网　　址　publish.tju.edu.cn
印　　刷　北京利丰雅高长城印刷有限公司
经　　销　全国各地新华书店
开　　本　190mm×219mm
印　　张　17.5
字　　数　234千
版　　次　2018年1月第1版
印　　次　2018年1月第1次
定　　价　188.00元

谨以此书

献给十年来砥砺前行的同事们

献给对北京建院约翰马丁国际建筑设计有限公司给予支持、关心和帮助的朋友们

序 · 至精至诚 品质至上

北京建院约翰马丁国际建筑设计有限公司（简称“建院马丁公司”）是北京市建筑设计研究院有限公司的国际子品牌，真正关注他们的作品是在上海世博会 BIAD 项目展上。在 2008 年举行的上海世博会沙特馆设计竞赛中，从来自全球的 30 个方案中选出 3 个方案进入第二轮，建院马丁公司提交的两个方案均进入前三名。虽然在最后的竞争中，建院马丁公司未能取得最终的设计权，但他们完成的设计方案仍然体现了相当的创新能力和设计水准，表现出他们对建筑与环境关系的理解以及对空间、构成的探索研究。

从上海 2010 世博会、西安 2011 世园会、北京 2019 世园会，到北京 2022 冬奥会……在祖国大地和国外的许许多多的项目现场都能看见他们团队勤劳的身影。仅几年时间，建院马丁公司涉猎了包括总部建筑、办公建筑、研发建筑、观演建筑、酒店建筑、医疗建筑、体育建筑、博览建筑、商业综合体、住宅、保障房建筑等类型，其中不乏超高层建筑。特别值得一提的是，这个团队在城市规划方面也作出了积极的探索。

经过几年的艰辛努力，建院马丁公司完成的设计作品开始出现在 BIAD 集团公司优秀工程评选中。下属公司不同于本部，优秀工程均是全部完成之后直接申报的，在设计过程中经常“犹抱琵琶半遮面”，所以经常会带来惊喜。例如，他们原创的吉安文化艺术中心、某研发中心项目等都给我留下了深刻的印象。某研发中心项目，虽然只有 6000 平方米，但这个公司在刚刚起步阶段尚面临生存与发展之矛盾时，仍能创作出与环境协调又构思独到的佳作，让我感到欣喜。吉安文化艺术中心刚刚建成就获得了“江西省首届十佳建筑”称号。

今年我受通用电气（GE）公司和建设方坤鼎集团邀请，去参观了建院马丁公司独立完成的新作“GE 北京科技园”。我清楚地记得四年前，建院马丁公司朱颖董事长向我汇报工作时讲道：“GE 是大型全球跨国企业，GE 北京科技园的设计对于团队是特别大的挑战。”GE 北京科技园全面打破格子间，构建了全新的共享、智能、社区化、园区化总部，令人耳目一新，给我留下了深刻的印象，即使作为 BIAD 本部，亦是一次完美的创新型尝试。项目后来获得了“全国勘察设计行业建筑工程一等奖”，可谓实至名归，这也是对建院马丁公司团队这些年辛勤工作的充分肯定。我们看到了“建院马丁”的品牌得到了越来越广泛的行业认知，也看到了建院马丁公司对于 BIAD 品牌建设方面所作出的贡献。

北京市建筑设计研究院有限公司作为与共和国同

龄的大型民用建筑设计企业，经过多年的积淀和历史传承，成为我国最有价值的建筑设计品牌之一，2017年更是获得了“亚洲最受尊敬的知识型组织大奖”。

几十年中，我们通过建筑设计服务于我们的国家、我们的人民。我们每个人都努力去做一名有职业素养的、有专业技能的、有社会责任感的、有人文情怀的建筑师。无论是过去、现在还是未来，BIAD 作为建筑设计行业的探索者始终秉承着对社会的尊重、对行业的尊重、对未来发展的尊重。

我们坚持职业、专注、协同、完善、创新的核心价值观，将建筑设计事业进行到底。

也正因为如此，作为唯一一家与集团本部同城同业的建筑设计控股子公司，建院马丁公司的发展和品牌建设之路并不平坦。他们成长的每一步，不但面临着激烈的外部竞争，同时无疑还有来自 BIAD 内部的激烈竞争。从金田公司的重组，到建院马丁公司发展的每一步，建院马丁公司的成长充满着勇于挑战的精神。

作为 BIAD 集团化战略的试点，我和朱小地先生一直有一个共同的原则，那就是“授之以鱼不如授之以渔”。我们更愿意让他们直接参与竞争，特别是面对社会上竞争激烈的市场环境，而不是直接给予经济上的扶植，因为只有经过真正的角逐才能够成长发展。建院马丁公司的大部分项目都是通过竞赛取得的，或者是原有客户的延续，这样当他们遇到困难时亦有顽强的生命力。在近年建筑市场受到冲击时，我们看到了这个团队知难而上的品格，团队领导班子敬业勤奋，创新自强；全体员工努力向前，不断探索。

也正因为如此，我愿意给予这个团队赞赏，赞赏他们迎接挑战的精神，他们的创新精神，他们对于建筑设计的热爱，他们内在的坚韧力量，还要赞赏他们边思考边设计的精神，我更赞赏这个团队至精至诚、品质至上的设计守则。

愿新的一年，能够有新的梦想实现。

北京市建筑设计研究院有限公司党委书记、董事长

2017 年 12 月 28 日

Preamble · Fineness and Sincerity, Quality First

BIAD John Martin International Architectural Design Co., Ltd. (BIAD–JAMA) is an international subsidiary brand of Beijing Institute of Architectural Design (Group) Co. Ltd. It was at the BIAD Project Show of Shanghai World Expo that their works were truly brought to attention. At the Saudi Pavilion Design Competition of Shanghai World Expo held in the year of 2008, three programs were shortlisted for the second round from 30 ones submitted from all over the world, and two programs submitted by BIAD–JAMA were shortlisted top three for the second round. Although BIAD–JAMA failed to obtain the final design right in the final competition, the design program completed by them still reflected the considerable innovation and design standard, showing their understanding of the relationship between architecture and the environment, as well as the exploration and research on the space and constitution.

From the 2016 Shanghai World Expo, 2011 Xi'an International Horticultural Exposition, 2019 Beijing International Horticultural Exposition, to Beijing 2022 Winter Olympics... their hard–working team can be seen at the scene of many projects in the motherland and abroad. In just a few years, BIAD–JAMA has been involved in a variety of architectural design types, including headquarters buildings, office buildings, research and development buildings, performance buildings, hotel buildings, medical buildings, sports buildings, expo buildings, commercial complexes, residential and security housing, and other types, some of which are super–rise buildings. The special team has made positive exploration in urban planning.

After several years of arduous efforts, the design works completed by BIAD–JAMA began to be shortlisted in the excellent projects of BIAD Group. Since the subordinate companies are different from the headquarters, the outstanding works are all declared directly after the completion, which always bring us surprises because they are not declared in advance during the design process. For example, the Ji'an Cultural and Art Center, and a Research Center originally created by themselves have made deep impression on me. Although the Research Center only covers an area of 6,000 square meters, I am very delighted that the company can still create such excellent works in harmony with the environment and with unique conception when the company is confronted with the contradiction between survival and development in the initial stage. The Ji'an Cultural and Art Center won the "First Session of Jiangxi Provincial Top Ten Buildings" upon completion.

I was invited by General Electric (GE) and the construction party Kuntin Group this year to visit the new works of "GE Beijing Technology Park (BTP)" independently completed by BIAD-JAMA . I remember that, four years ago, Zhu Ying, the Chairman of BIAD-JAMA, said when reporting the work to me, "GE is a large-scale global multinational corporation, so the design of BTP is a great challenge for the team." The completed BTP completely breaks out the layout of cubicles, constructs a whole new world of sharing, intelligence, community-based and park-based headquarters that makes people refreshed, leaving a deep impression on me. Even for the headquarters of BIAD, it is also a perfect innovative attempt. The project won the "First Prize of National Survey and Design Industry Construction" later, which deserved its full recognition and gave full affirmation to the team of BIAD-JAMA for the hard work of these years. We have seen more and more extensive industry recognition that the brand of "BIAD-JAMA" has won, especially the contribution made by BIAD-JAMA on the brand construction of BIAD.

As a large-scale civil architectural design enterprise with the same age with the People's Republic of China, BIAD has become one of the most valuable architectural design brands in China after many years of accumulation and historical heritage. In 2017, it won "Asia's Most Respected Knowledge-Oriented Organization Award".

For several decades, we have been serving our country and people through architectural design. We strive to be a professionally qualified, occupationally skilled, socially responsible, and humane architect. Whether in the past, present or future, as the explorer of the architectural design industry, BIAD always adheres to the respect for the society, industry and the future development.

We adhere to the core values of occupation, focus, coordination, improvement and innovation to carry out the cause of architectural design through to the end.

Precisely because of this, as the only architectural design holding subsidiary in the same city and the same industry with the headquarters of the Group, the development and brand building road of BIAD-JAMA is not flat. Every step of its growth is not only faced with fierce external competition, but also undoubtedly faced with intense competition within BIAD. From the reorganization of Jintian Company, to each step of the development of BIAD John Martin, the growth of BIAD-JAMA is full of the spirit of daring to challenge.

As a strategic pilot of BIAD Group, Mr. Zhu Xiaodi and I have a common principle, namely "teaching them how to fish rather than giving them fish". We are more willing to let them take part in the competition directly. Especially in the face of the fierce competitive market environment in the society, they can grow up and develop only with the real competition rather than the direct economic support. Most of the projects of BIAD-JAMA were won through competitions or the continuation of existing customers, so that they have tenacious vitality when they encounter difficulties. We have witnessed the quality of hard working in face of difficulties of the team when confronting the impact of the architecture market in recent years. The team leadership is dedicated, hardworking, and innovative with self-improvement; and all the employees forge ahead with constant exploration.

Just because of this, I'd like to give this team appreciation for their adventurous spirit, innovative spirit, their love for architectural design, their inherent strength, as well as the appreciation for their spirit of designing amidst thinking. I also appreciate the team for the design principles of fineness and sincerity, and quality first.

I wish new dreams will come true in the next year.

Xu Quansheng
Party Secretary and Chairman of
Beijing Institute of Architectural Design
(Group) Co., Ltd.
December 28, 2017

序 · 《建筑拾记》读后

北京建院约翰马丁国际建筑设计有限公司（简称“建院马丁公司”）成立于 1994 年，那时称为北京金田建筑设计有限公司，2007 年北京市建筑设计研究院对金田公司进行了战略改组， 2010 年引进了美国的约翰马丁结构工程设计集团作为战略投资人，并正式更改为现在的名字，目前由青年建筑师朱颖出任董事长兼总经理。其实对美国的约翰马丁公司我还是有点儿了解的，1984 年 8 月去考察美国体育建筑时，曾在洛杉矶顺访过美国约翰马丁公司，并与公司负责中国业务的罗超英女士有过几次会面。美国的约翰马丁公司以结构设计见长，两家合作以后，以北京建院为技术支撑，成为由中华人民共和国商务部批准，住建部颁发甲级工程设计资质证书的中外合作企业，也是北京建院在深化改革、探索创新道路上一次重要的尝试。公司改组 10 年来，完成了近 200 项、近 1000 万平方米的建筑设计及近 100 项规划设计项目。公司致力于提供从建筑策划到规划方案、施工图直到后期服务的全过程综合服务，承接了国内外的一批重点项目，获得了许多国家和地方的重要奖项，成绩斐然，展现了整个设计团队的综合实力和管理水平。最近朱颖将他们团队改组 10 年来在各设计项目中的体会和感悟总结为《建筑拾记》一书，我有幸得以先睹，受到很多启发。

建筑设计活动是把知识和技术转化为现实生产力过程中的关键环节，在设计的全过程中，建筑师的思路、感悟、知识、意志、经验、价值观和审美观都是一次充分的体现和表达，表现了设计思维活动的极端复杂性。这里面表现出对严密的逻辑性和艺术个性的统一，对安全性和风险性的认知，对技术因素和价值因素的追求等过程。而另一方面设计过程的创新不是基础科学层面的原始创新，而是通过知识和技术的渐进性积累、综合集成而逐步完善和推进。因此设计活动的理性思维对于建筑师的成长和设计团队作品的质量十分重要。朱颖和他的设计团队总结说：“团队一直在探索和思索中前行，自成立以来，对公司的环境观、城市观的探索，对文化引领设计手法的研究，对新建筑技术的研发，对团队设计观的总结一直都在进行中。”这与在当前城市化的大潮中，许多建筑师满足于“作而不述”形成了明显的对比。《建筑拾记》涉及了对不同建筑类型和设计项目在不同层面和不同学科维度上的思考和提升，相信对设计团队的提高甚至是行业的技术进步都是十分有用的。我想以“拾记”中的两例来谈谈自己的体会。

“壹记”的题目是“未来园区办公”，这是美国通用电气（GE）公司在北京科技园的跨国研发企业总部设计。GE 是全球最大的跨行业经营的科技、制造和服务型企业之一，世界五百强排名第 16 位。这次公司把其在中国的运营管理、研发设计及部分销售整合在一起，打造 GE 北京科技园。其总用地面积 5 公顷，总建筑面积 7.4 万平方米，地上 5 层，地下 1 层。项目于 2016 年 10 月正式落成启用，该设计项目获得了 2017 年全国优秀工程勘察设计行业奖的一等奖。

近代办公建筑的发展从 20 世纪初至今，也就有百余年的历史，但办公建筑的出现被认为是 20 世纪最重要的标志物之一，是社会进步、技术发展、经济繁荣的产物。从 1885 年美国芝加哥第一栋 10 层的报销公司大楼开始，高层商务办公建筑作为一种新的建筑类型随着电梯的出现而华丽登场，芝加哥学派也在高层建筑的发展史上占有重要地位。此后建筑的重心转向纽约；折中主义风格成为典型的设计风格，高度不断被突破，建筑材料、施工机械、新型设备等技术进步为这些突破提供了可能。办公建筑的组织管理、研究决策、信息处理功能也更加高效和细化，并很快转入了国际式的现代风格，尤其到 20 世纪 70 年代前后进入高层办公建筑的极盛时期。提高办公效率、营造符合心理和生理需求的工作环境，控制建造和运营成本以及有特色的建筑形象甚至公共利益的考虑成为新型办公建筑的基本设计目标。尤其是随着计算机技术、网络信息技术的进步，许多高科技产业、创意产业、制造产业出现了总部办公式研发型的办公建筑。除了其企业管理、厂品展示、企业宣传、公共开放等特点外，该类建筑更加注重激发创造性、绿色生态环境、安全私密性、人性化关怀、员工交流等方面的追求。通用公司的研发企业总部就面临着这样的新挑战、新要求和新目标。

朱颖和他的设计团队为了应付这一挑战，进行了对研发型总部办公设计的调研和案例分析。他们通过国内外对苹果公司总部、谷歌公司总部、Face Book 总部、蚂蚁金服总部大楼的分析和解剖，总结出了这种面向未来的研发办公的一些基本特色和趋向，诸如：扁平的组织管理结构激发灵感，促进沟通；办公环境的可改变的弹性以创造共享、分享、交换和交流；智能与物联网 + 傻瓜式体验等健康办公理念，建立起迅速有效的信息传输机制；轻量化社区 + 管家式模式与综合管理，形成服务和生活配套设施的诸多空间……这样为设计团队的思路和设想提供了切实可行的目标，进而总结出“理性创新、面向未来、绿色生态”的设计原则，旨在设计一个符合 GE 公司全球化企业形象的创新型总部，富有归属感的企业园区，同时它也将成为孕育梦想的创新场所。

正是按照这样的思路，朱颖和他的设计团队一起在科技园的办公空间构成上，首先体现扁平化管理形成的平等与交流的组织管理体系，传递轻松而自由的交流氛围；其次主要营造一个功能齐全、设备多样的生态办公社区，通过提供工作、就餐、休息、培训、娱乐、健身、医疗等方面的设施形成员工新的生活方式；另外设置了共享、可变可拓展的办公空间，通过网络地板便于工作的即时更新；在设施管理上通过智能安全管理，智能工位预留系统、智能会议预留系统、智能楼宇控制等实施国际化的管理服务，使科技园通过了美国 LEED 绿色建筑认证，同时通过了 FM 全程的安全性能评估，使建筑在运行费用、节能减排、创造舒适和人性化的工作环境上，特别是在建筑安全性上均达到一定水准。因此，设计团队认为总部办公空间与一般的办公建筑不同，当前加强这方面的研究和探索，将更加有助于加快我国科技创新和赶超国际先进水平的步伐。

“叁记”的题目是“城市活力客厅”，介绍了江西吉安文化艺术中心的设计。这是一个以 1157 座席的剧院为主体，集观演、会议、展示、休闲、办公为一体的大型综合文化建筑。2012 年项目建成后获江西省首届十佳建筑奖，2015 年获全国优秀工程勘察设计行业奖二等奖。除了在设计过程中对城市空间的整合、整体布局和造型上的经营工作外，团队对于剧场的托管经营和后评估工作也别具特色。

近年来我国在剧场类文化设施的建设在各地掀起一股热潮，据统计到 2016 年年末我国共有艺术表演场馆 2285 个，比 2015 年增加 142 个。但据文化部的统计，各级文化部门所属场馆年均演出仅 43 场，每个观众席年利用率仅 26.5 人次，相当于全年有 338 天场馆处于空闲状态。2015 年全年全国艺术表演团体共演出 210.78 万场，但进入剧场的仅有 9 万场；全年有 9.58 亿人次观看演出，但进入剧场的观众仅有 4000 万人次。这说明剧场行业在设施利用上存在巨大的浪费，设施长期空置，有巨大的提升潜力。因此，在对剧场经营管理模式进行优化改革的同时，对已建剧场设置的评估和分析工作显得十分必要和紧迫。

当前“前策划后评估”的知识体系和设计方法已越来越为人们所重视。国外在 20 世纪随着控制论的出现，通过数据的聚集、信息反馈来研究系统的控制和调解，以解释其中的控制规律。对于后评估（POE）工作，英国皇家建筑师学会曾定义为：“建筑在使用过程中，对建筑设计进行的系统研究，从而为建筑师提供设计的反馈信息，同时也提供给建筑管理者和使用者一个好的建筑标准。”我国张钦楠先生曾在1995年较早地介绍过这一方法。近年来住建部在2014年提出“探索研究大型公共建筑的后评估”，2016国务院提出“建立大型公共建筑工程后评估制度”。“前策划后评估”形成一个闭合环节，但在性质上还是有所区别的，前策划更侧重于设计工作前“解决有无”的界定计划，而后评估则着重于建设成果好与坏的反思和评价服务。比较之下，后者的难度更大、准确度要求更高。

吉安文化艺术中心的设计过程充分考虑了这一问题。早在设计之初市委市政府就决定设施建成后将由北京保利剧院管理有限公司托管经营，该公司在全国各地托管了53家大剧院，在技术、管理、运行、保障诸方面都很有经验。根据托管方的意见，设计团队在设计之初就对设计内容进行了调整和修正。在建成以后的托管过程中，根据“市场运作、业主监管、委托经营、政府补贴”的经营管理模式，运行十分顺利。在2017年托管合同5年期满时对运行情况进行了后评估，据统计5年托管期中自营组织演出165场（平均每年30场以上），本地演出235场，另外每年有地方会议20场、市民活动12场，接待中央和外省市参观考察团年均46次，每年的复合使用天数为132天，重要设备完好率常年达到98%，5年间从未发生重大安全和设备事故。在后评估的基础上，决定继续委托保利公司托管到2022年。由于关注了建筑生命周期的全过程，为设计建成后的运营管理创造了良好的条件。这种前策划后评估的体制，考验了决策的科学性，实施的完整性和可持续性，不仅是设计个案提升运营潜力、改进工作的重要反馈和指导，对于行业的健康发展也具有指导意义。

朱颖和他的设计团队总结的《建筑拾记》还有许多亮点和真知灼见，对于充满激情的创作团队而言，10年还是一个不长的周期，还有较长的路要去探索。建院马丁公司本着积极求索的初衷，不断开拓进取，“不积跬步无以至千里，不积小流无以成江海”，已经取得了十分可喜的成绩。仅以这短短的读后记表示对朱颖和他的团队的祝贺和期望。

中国工程院院士
全国工程勘察设计大师
北京市建筑设计研究院有限公司总建筑师
2017年12月23日

Preamble · Review of the Book *Shi Notes on Architecture*

BIAD John Martin International Architectural Design Co., Ltd. (the Company) was founded in 1994. Back then its name was Beijing Jintian Architectural Design Co., Ltd. In 2007, Beijing Institute of Architectural Design (BIAD) took the lead in strategically restructuring it with young architect Zhu Ying serving as its chairman and general manager. John A. Martin & Associates, Inc. (JAMA), a US-based company, was introduced as a strategic investor in 2010, since than the Company changed into its current name. JAMA is not a strange name to me. Back in August 1984, I ever visited the US for inspecting some of its sports buildings. During my tour, I dropped by JAMA, and met with Ms. Luo Chaoying in charge of the company's business in China for several times. JAMA is famous for its structural design and BIAD is able to provide strong technical support. The partnership of the two gives birth to a Chinese-foreign joint venture which is approved by the Ministry of Commerce of the People's Republic of China and licensed by the Ministry of Housing and Urban-Rural Development of the People's Republic of China as a grade-An engineering design agency. At the same time, it also represents an important attempt made by BIAD to go deeper in reform and innovation. Over the past decade since the reorganization, the Company has completed 200 architectural design projects involving nearly 10,000,000 m^2, and 100 architectural planning projects. The Company is dedicated to providing the comprehensive architectural services that cover the entire construction flow: architectural proposal, planning scheme, working plan, and post-occupancy services. It has contracted a raft of key projects at home and abroad, and won an abundance of great prizes at national and local levels. What the Company has achieved is a convincing testimony to the comprehensive strength and management capacity of its entire design team. Recently, the General Manager Zhu Ying compiled the practices and reflections his team had made in the past ten years' work into a book named *Shi Notes on Architecture*. I have been honored to read it first, and it is a thought-provoking volume. The following are some inspirations the great book has given to me.

Architectural design constitutes a key link in the effort to convert the knowledge and technology into the practical productive forces. In the design process, architects are well positioned to fully express their ideas, thoughts, expertise, willingness, experience, values and aesthetics.

In this sense, design involves the extremely complicated conceptual work. On the one hand, architectural design comes as a process that combines the rigid logics and the distinctively artistic personality of designers. It represents the ways they perceive the relationship between security and risk, and balance their pursuit for technical and value factors. On the other hand, to make innovation in the design process, architects usually need to gather, integrate, and refine more expertise and techniques step by step, rather than create primitively new things on the level of basic science directly. Therefore, rational thinking in design activities is of critical importance for both the career development of architects and the production of high-quality works for design teams. "We have always been feeling our way as we go. Since the very beginning, we have made explorations to forge our unique corporate views about environment and cities , studied how to base our design techniques on culture, and developed new architectural technologies, and summarized what our team achieved into design concepts", as General Manager Zhu and his team summed up. This move makes a striking contrast with many other architects who are confined to the sheer practice amid the current urbanization drive. *Shi Notes on Architecture* is a collection of architects' reflections on and improvements in various types of buildings and design projects on different levels and across diverse disciplines. I'm convinced that it can help design teams to hone their skills and even propel the entire industry towards more progress. In the following part, I would like to share my views about the two projects extracted from the ten ones of the book.

Note one comes under the title *Future of Workplace*. It is about designing the transnational corporate R&D headquarters for General Electric (GE) in Beijing Science Park. General Electric is one of the world's largest companies that operate across multiple sectors including technologies, manufacturing and service delivery. It is No.16 on the Fortune 500 list. Through this project, GE intended to bring together its operational management, research, development and design, and sale (partial) resources in China, by building GE Beijing Technology Park (BTP) in the country. The project covers a land of 5 hectares and has an overall floorage of 74,000 square meters, with five storeys above the ground and one underground. It was completed for use in October 2016. The design project won the first prize of the "Exemplary

Engineering Prospecting and Design Awards" in 2017.

It only takes 100-plus years for the modern office building sector to get where it is now from its advent in the early 1900s. Office building is deemed as one of the most important symbols of the 20th century. They are the results of social progress, technological development, and economic flourish. It is in 1885 that Chicago erected its first ten-storey Home Insurance Building. Since that, the high-rise business and office buildings have made their striking appearance as a new construction type with the invention of elevators. The Chicago School thus occupied a significant position in the course of development of high-rise office buildings. After that, the focus of the building industry started switching to New York. Eclecticism became the typical design style. New records of building height are being created, which is partially credited to the progress in building materials, construction machinery, new-type facilities, and other aspects. At the same time, office buildings demonstrated more efficient and meticulous organizational management, research and decision-making, and information processing functions. In no time the international modern style came into being. Especially, the high-rise office buildings came in full swing around the 1970s. Almost all designers, while planning new-type office buildings, intend to improve working efficiency, create workspaces that carter to people's psychological and physiological needs, control construction and operational costs, shape unique building images, and even take care of public interests. Particularly with the development of the computer, Internet, and information technology, many hi-tech industries, creative industries, and manufacturing industries erected the headquarters-based R&D office buildings. While integrating the functions of corporate governance, product presentation, company publicity, and openness to the public, the new-type facilities weigh more on the pursuits for creativity cultivation, eco-friendliness, security and privacy, fulfillment of personalized needs, and efficient communication, among other things. These are the new challenges, requirements and objectives confronting the current construction projects. GE's corporate R&D headquarters is no exception.

To rise to the challenges brought by the GE project, Zhu Ying and his team surveyed and analyzed many corporate R&D headquarters as cases. A close look at the headquarters buildings of many companies at home and abroad including Apple, Google, Facebook, and Ant Financial awakens them to know some features and trends the future-embracing R&D workspaces share in common. As they observed, these office buildings tend to adopt a flat organizational structure which helps to spark inspirations and encourage communications; the workspaces are usually flexible and changeable so that people working there can share their opinions as they want; the introduction of such sound working concepts like artificial intelligence, Internet Plus and user-friendly experience enable these workspaces to adopt the efficient information transmission mechanism; and the design of lightweight-community and house-keeping mode and comprehensive management helps to create many spaces with various service and living facilities available... By sorting through their ideas and assumptions, the design team worked out a feasible objective and then came up with a principle of "making rationally innovative, future-embracing and eco-friendly design". As they envisioned, there would be a creative headquarters in line with GE's corporate image as a global conglomerate corporation, a corporate park making people feel at home, and also an innovation incubator where dreams come true.

Tracking the train of thought, Zhu Ying and his design team proposed the following workspace layout. First of all, a flat organizational structure was adopted to advocate equality and face-to-face communication, to create a relaxing, free working environment. Secondly, there emerged a fully-functional, well-equipped eco-office community where employees could enjoy a new lifestyle that integrates work, dinning, rest, training, entertainment, fitness, medical care, and activities all together. Thirdly, shared, changeable and expandable workspace was created by applying the network floor system which could be updated instantly as needed by the actual work. Fourthly, the intelligent security management system, the intelligent workstation reservation system, the intelligent meeting reservation system, and the intelligent building control system were employed together to manage the work places up to the international standards. The science park has been recognized as a green building complex by the Leadership in Energy and Environmental Design (LEED) and it has passed the full FM safety performance assessment, which signaled that it attained specified achievements in terms of operational cost, energy conservation and emission reduction, comfortable and user-friendly working environment, and building safety in particular. Therefore, the design team holds that the workspace is different from the general office buildings. As it suggests, the researches and explorations in this regard need to be stepped up now. By doing so, China can gather its pace in making technological innovation and catching up with leading competitors in the international community.

Note Three comes under the title *Reception Room of Urban Vitality*. It introduces the design of Ji'an Cultural and Art Center in Jiangxi. Ji'an Cultural and Art Center is a versatile cultural building that integrates performance view, meeting, presentation, leisure, workspace and other purposes. At its center stands a theater which can hold 1,157 people at most. Right after its completion in 2012, the structure was named one of the "Best Ten Buildings of Jiangxi Province". In 2015, it won the second prize of the "Exemplary

Engineering Prospecting and Design Awards". While planning the center, designers took into account the spatial features of the city, and thought hard about the overall structure and shape of the building. Besides, they also worked out a unique plan about the trusteeship management and post-occupancy evaluation of the theater.

In recent years, cultural facilities represented by theaters have sprung up one after another all over China. According to the statistics, China had 2,285 art performing venues in total by the end of 2016, up by 142 over 2015. But the data released by the Ministry of Culture suggested that the annual average number of performances hosted by the venues affiliated to cultural authorities at different levels was merely 43, and each auditorium received 265,000 spectators every year averagely, which means that these venues remained idle for 338 days per year. In 2015, art performance troupes delivered 2,107,800 performances, out of which only 90,000 occurred in theaters. Throughout the year, 958 million spectators watched these performances, of which only 40 million went into theaters. The huge gap indicates that the theater facilities nationwide are extremely under-utilized and stay idle for the jaw-dropping period of time. There is huge room for improvement. In addition to the change and improvement in the theater operating mode, it is also quite imperative to evaluate and analyze how the existing theaters are set up.

Now the "preplanning and post-occupancy evaluation (POE)" knowledge system and design method is drawing more attention with each passing day. With the advent of control theory in last century, the international community began studying how to control and mediate various systems and explain the control laws contained thereof through data collection and information feedback. Post-occupancy evaluation is "a systematic study of buildings in use to provide architects with information about the performance of their designs and building owners and users with guidelines to achieve the best out of what they already have" as defined by the Royal Institute of British Architects (RIBA). Back in 1995, Chinese scholar Zhang Qinnan introduced the concept to China. In recent years, the Ministry of Housing and Urban-Rural Development proposed to explore and study the POE of large-sized public buildings in 2014, and the State Council asked to put the large-sized public building projects under the POE system. Pre-planning and POE form a closed loop. But the two are different in essence. Pre-planning is focused to define whether a design problem is solved or not while POE is primarily intended to reflect and evaluate the effects of design results. The comparison between the two tells us the latter is more demanding in complexity and accuracy than the former.

The design of Ji'an Cultural and Art Center took POE into full account. At the very beginning of the design, the municipal party committee and the municipal government decided to commission the

Beijing-based Poly Theater Management Co., Ltd. (Poly) to manage the center upon its completion. The company now runs a total of 53 grand theaters nationwide on commission. It is highly experienced in related technology, management, operation, and support. According to the opinions given by the trustee, adjustments and modifications were made to the design from the very start. The center has been run very well by Poly. Following the market laws, its operation is supervised by the proprietor and funded by the subsidies from the government. The POE was given to the center in 2017 after the five-year trusteeship contract expired. According to the statistics, the center staged 165 performances on its own (more than 30 performances per year), and hosted 235 local performances in the past five years. Besides, it also convened 20 local meetings, held 12 citizen activities, and received 46 delegations from the central government and other provinces/ municipalities averagely each of the five years. Over the measured period of time, its facilities were put under compound use for 132 days per year averagely, and the rate of major equipment in good conditions remained as high as 98% throughout the year. No material security incident or equipment failure occurred over the past five years. According to the POE findings, the center decided to extend its trusteeship contract with Poly till 2020. The concerns for the entire lifecycle of the building laid solid grounds for the post-occupancy operation and management. The mechanism combing pre-planning and POE is designed to test the rationality of the related decisions and the integrity and sustainability of the implementation. As a result, it serves as a significant feedback and guidance for individual design cases to improve operational potential and work efficiency, and also helps to promote the entire industry towards sound development.

Shi Notes on Architecture is a collection of wisdom of Zhu Ying and his design team. Due to limited space, I just name a few highlights and insights contained in the book here. For a passionate creation team, ten years is just a short period of time. Ahead it is a long way to go. Staying true to its initial aspiration, the Company will bear in mind what has been achieved and continue to forge ahead courageously. In closing, I would like to extend my congratulation and good wishes for Zhu Ying and his design team.

Ma Guoxin
Academician with the Chinese Academy of Engineering
National Master of Engineering Survey and Design
Chief Architect with BIAD
December 23, 2017

前言 · 初心秉持

您能读到这段文字，是我的至上荣幸，希望这本《建筑拾记》能让您了解这个执着的团队，或者给予赞许的目光，或者提出中肯的建议，抑或是结缘共同开始做一些对我们赖以生存的城市有意义的事情，即便只是让您了解了建筑背后的故事，我都已心满意足。

2007 年 10 月，根据北京市建筑设计研究院的指示，我被委派到北京金田建筑设计有限公司（2010 年更名为北京建院约翰马丁国际建筑设计有限公司）参与重组工作。这本《建筑拾记》记录了十年来我们在祖国大地完成的拙作，它们背后是大家倾心挥洒的辛勤汗水，是团队每位成员的初心秉持，它更是我本人及团队十年探索之路的精华萃取。

我认为，一个良好的创作观和利于创作的环境，是一个建筑设计团队的核心价值，我希望公司是个可以平等交流的地方，是个开放的空间。公司一直崇尚创新、尊重创新，截至 2017 年 11 月，我们共获得了包括中国建筑学会的建筑创作金奖、中国勘察设计行业协会优秀工程一等奖在内的各类奖项近 30 项，无疑这对于一个中型设计团队而言是极其难得的。

我始终认为，“设计是尊重内心的行为方式，始于心，而不止于心”。好的建筑设计是一种尊重，是建立在给予人、社会、文化、环境等客观条件以足够尊重的基础上，而产生的共生共存的协调关系，但好的建筑又在共生协调的基础上做一个舞者，谦逊但真实地表达自我，也许人、社会、文化、环境等互相尊重，也是为自己的设计得以更好实现的一种手段。

我们愿意把建筑视为一个生命体，设计只是一个孕育的开始，我们始终秉持“匠人精神”，希望能用坚实的技术功底使作品有更高的完成度，但我们同样关注建筑的易建性，让建筑能够在各方面客观条件不一定尽如人意时同样能完美地绽放。

我们更愿意去影响每一位项目的参与者，也许他是主管建设的决策者，也许他是投资人，也许他是承建方，也许他是供应商，也许他是运营团队，也许他是建筑的最终使用者，也许他和我们一样是建筑师……我们都希望能影响他热爱建筑，愿意和我们一起呵护建筑的成长。爱是建筑完美绽放的前提，我们关注它全生命周期中的每一个阶段，我们希望整个的生命周期中建筑都能得到很好的爱护。

我期待，我们的设计能给人带来美的感受，能给人带来愉悦。勒 · 柯布西耶曾说过：“你运用石头、木头和混凝土等材料盖出房舍和宫殿——这是建造。但是如果在建筑中再融入独创性和工艺，并且你忽然之间打动了我的心，让我感受到愉悦——我很快乐。然后我说，这真美。这才是‘建筑’。”

我更期待，我们的工作对于城市和环境产生积极的意义。梁思成先生说过：“城市是一门科学，它像人体一样有经络、脉搏、肌理，如果你不科学地对待它，它会生病的。”如果说用钢筋水泥筑成的建筑是城市的肌体，那么，附着在建筑之上的历史记忆、民族情感与人文精神则是城市的精气神。设计还要对生态文明如何发挥保护环境的引领作用，提供有价值的规划设计献策。建立超越一般设计师的城市观、文化观、环境观，一直是团队的目标和坚定自信的关键。

我希望用我们的理性思考向《建筑十书》的作者维特鲁威致敬，因他提出的坚固、实用、美观的建筑学理论，在今天仍可作为我们建筑师恪守的原则之一。理性之于一个设计团队，正如思想之于一个卓越的设计师和管理者：十年设计创新路，我们并非为理论抽象而思考；十年设计创新路，无论是前瞻性、规律性都是在实践背景下应运而生的；十年设计创新路，更告诉我们，建筑需要诚实，需要美的法则和以记忆为载体，无论是设计师还是管理者都要使自己成为那种内外感觉都协调一致的人，成为那种直至成年也依旧童心未泯的纯粹之人。

大道至简，平淡为归。现代主义建筑大师密斯·凡·德·罗曾有一句“少即是多”（Less is more.）的名言，而我们特别尊重的工业设计大师迪特·拉姆斯（Dieter Rams）却将他的设计理念阐述为“少，却更好”（Less, but better.）。

我们当年曾经反复吟读迪特·拉姆斯关于好的设计（Good Design）的十项原则，那时读来觉得颇为平淡无奇，而从业多年后回顾才发觉，这样解读了“基础的根源”的设计原则其实真是知易行难。

在最后，我愿意再次与大家分享这十项原则并与大家共勉：

好的设计是创新的；
好的设计是实用的；
好的设计是唯美的；
好的设计让产品说话；
好的设计是谦虚的；
好的设计是诚实的；
好的设计是坚固耐用的；
好的设计是细致的；
好的设计是环保的；
好的设计是极简的。

拾年韶华，我们从未辜负；前途漫漫，携手再赴新途。

朱颖
北京建院约翰马丁国际建筑设计有限公司
董事长、总经理
2017 年 10 月 24 日

Foreword · Defend the Beginning Mind

I feel so honored that you read this passage. I hope the *Shi Notes on Architecture* will let you know about our dedicated team, or give it a thumbs-up, or give some pertinent advice, or start to do something together for the sake of the cities we live in. Even if it is just for letting you know the story behind the buildings, I am satisfied with it.

In October 2007, according to the instructions of Beijing Institute of Architectural Design (BIAD), I was sent to the then Beijing Jintian Architectural Design Co., Ltd. and responsible for the reorganization work. A strategic investor was introduced in 2010 when Beijing Jintian Architectural Design Co., Ltd. changed its name to BIAD John Martin International Architectural Design Co., Ltd. (the Company). The *Shi Notes on Architecture* records what we have achieved in our motherland in the last 10 years. This book epitomizes the hard efforts of us, the support of every team member, and a selection of the results of our ten-year exploration.

In my opinion, a good view of creation and a creation-friendly environment are the core values of an architectural design team. I hope the Company is an open place where we can communicate with each other on an equal footing. The Company has been advocating innovation and respecting for innovation. As of November 2017, we have won a total of nearly 30 awards, including the Gold Award of Architectural Creation of Architectural Society of China, the First Prize of Excellent Engineering of China Engineering & Consulting Association. This is quite marvelous for a medium-sized design team.

I always think that "design is a way of act that respects for the heart, starting from the heart but beyond the heart". Good architectural design is a form of respect and a coordinated relationship of coexistence resulting from the sufficient respect given to objective conditions such as human, society, culture and environment. However, a good building is like a dancer based on the coordinated relationship of coexistence, expressing itself humbly and authentically. Design may originally be selfish, but can be better realized through mutual respect of human, society, culture and environment.

We are willing to regard a building as a living body. We hope that the design represents only a start and can be better realized in the construction process. We always uphold the spirit of "craftsmanship" and hope to complete our works at a higher level with a solid technical foundation. In the meantime, we pay attention to the ease of construction of buildings, so that buildings can bloom excellently when the objective conditions in various aspects are not always satisfactory.

We are more willing to influence every participant in a project, be it a decision maker in charge of construction, an investor, a project manager, a contractor, a supplier, an operation team, an end user, or an architect like me... We hope that every participant could love the industry, love architecture, and be willing to care for the growth of architecture. The love of all people underpins the perfect bloom of architecture. We pay attention to every stage in the whole life cycle of architecture. We hope that architecture can be well cared for in its entire life cycle.
I expect that our design can bring people the feeling of beauty and joy. Le Corbusier mentioned in his book *Towards a New Architecture* published in 1923, "You employ stone, wood and concrete, and with these materials you build houses and palaces. That is construction. Ingenuity is at work. But suddenly you touch my heart, you do me good, I am happy and I say, 'This is beautiful.' That is Architecture. Art enters in."

I further expect that our occupation has a positive meaning for cities and environment. Mr. Liang Sicheng said, "The city is a science. It has meridians, pulses, and textures as the human body. If you treat it unscientifically, it will get sick." If buildings made of reinforced concrete are the body of a city, then historical memories, national emotions and cultural spirits attached to the buildings are the essence of the city. Culture is the soul of urban architecture and an important manifestation of urban soft power. Design should provide new valuable planning and design ideas as to how to play the leading role of ecological civilization in protecting the environment. At the same time, a view of design takes full account of the big-picture planning and environment at the regional level. Designing urban architecture with distinctive characteristics, profound cultural background, harmonious and unified styles, and big picture of landscape environment has always been the highest goal pursued by our team. Establishing a view of city, a view of culture, and a view of environment beyond average designers is the key to the team's firm confidence.

I would like to pay homage to Vitruvius, the author of the *Ten Books on Architecture*, with our rational thoughts. His theory of firm, practical and beautiful architecture is still one of the principles our architects abide by today and is a basic standard we are trying to practice. Rationality is important to a design team as if ideology is important to a brilliant designer and manager. Over the ten years of design innovation, we have pursued creative thoughts not for theoretical abstraction; over the ten years of design innovation, our forward-looking and regular ideas have come into being in the context of practice; over the ten years of design innovation, we have learnt architecture needs honesty, authenticity, law of beauty, and memory as a carrier. Both designers and managers must make themselves the kind of people who feel coordinated internally and externally and have pure souls who retain childlike innocence.

The greatest truth is the simplest. "Less is more" is a famous saying of the modernist architect Mies van der Rohe, while the esteemed industrial designer Dieter Rams put his design philosophy as "less, but better".

We once read quite a bit about Dieter Rams' ten principles of good design and felt rather insipid at that time. After many years of practice, we find that those design principles interpreting the "source" of design basis are said more easily than done.

In the end, I would like to share it with everyone again and encourage each other:

Good design is innovative;
Good design makes a product useful;
Good design is aesthetic;
Good design helps a product to be understood;
Good design is unobtrusive;

Good design is honest;
Good design is durable;
Good design is thorough to the last detail;
Good design is concerned with the environment;
Good design is as little design as possible.

We have lived up to the past ten years and we would like to join hands with you to embark on a new journey.

Zhu Ying
Chairman and General Manager of BIAD John Martin International Architectural Design Co., Ltd.
October 24, 2017

目录

Contents

壹记 Note One

未来园区办公
Future of Workspace

GE北京科技园设计与园区型总部发展探讨
GE Beijing Technology Park (BTP) Design and Corporate Headquarters Evolution

互联网和第四次工业化浪潮，对传统办公模式和空间提出了挑战，
互联网、视频设备的普及，使办公场所的重要性显著降低，
办公空间存在的意义受到质疑。
自从互联网兴起以来，世界以前所未有的速度掀起了商业变革，
过去需要几十年甚至上百年才能打造出来的商业巨头，
现在可能只需要几年的时间，
传统的跨国企业在新的全球互联背景下同样开启了迅速扩张之路，
对跨国型企业总部园区建筑设计同样提出了挑战。
全面取消格子间，智能互联、绿色健康的理念将引导办公空间的未来。

The Internet and the Fourth Industrial Revolution have presented challenges on the traditional work style and office space. The pervasive Internet access and the video equipment make the workspace much less important than it was. And even the existence of office space comes under scrutiny. The rise of the Internet has ushered in an era of unprecedented business revolution around the world. Now it only needs a few years for a business magnate to grow from scratch. But in the past, it used to take several decades or even one entire century to do so. In the meantime, the traditional transnational corporations also embark on a path of fast expansion in the new round of globalization drive, which also brings forward challenges to the architectural design of the headquarters parks of transnational corporations. The future office space will be led by the concepts of overall eradication of cubicles, intelligent interconnection and green healthy development.

一、传统办公空间遇到的挑战与园区型总部建设热潮

兴起于20世纪60年代的格子间式办公，曾是办公空间的主角，到90年代，开放式办公开始大规模出现，2000年以后，硅谷新兴的高科技公司开始引领开放式办公的潮流。近年来，移动互联、无线网络、互联网、视频设备的普及，使办公的场所不再那么重要，只要有网络，职员随时随地就可以完成工作。这对传统办公空间和办公模式均提出了挑战，办公空间存在的意义受到质疑。我们甚至可以看到，小微公司的SOHO办公、“咖啡”办公、双创空间、共享办公（Co-work）等开始大量出现，并造成传统的大型写字楼空置。

在传统办公空间遭到质疑的同时，我们也看到，大型跨国企业投入巨资打造新型园区总部却正在成为一个新潮流。例如，苹果公司耗资50亿美元建设的总部园区(Apple Park)将于近日正式开放，亚马逊在西雅图建设的新总部，Frank Gehry操刀的Facebook新总部园区，BjarkeIngels（BIG）在负责设计谷歌(Google)在加州的新总部“山景城”，在国内，百度、京东、腾讯、阿里巴巴、苏宁、小米等都在近年建设或者准备建设新的总部园区。

可以确定的是，传统的办公空间和模式不能再满

图 1-1 西北侧入口实景

足这些跨国公司总部的需求，它们需要符合自身发展的新型办公空间。一方面，以互联网产业化、工业智能化、工业一体化为代表的第四次工业革命使企业创新和颠覆的速度极大加快，过去需要几十年甚至上百年才能打造出来的商业巨头，现在可能只需要几年的时间，“独角兽”企业大量出现，在中国以百度、阿里巴巴、腾讯为代表的互联网企业同样发展迅猛。另一方面，传统的跨国企业在新的全球互联背景下同样开启了在传统领域内继续引领占全球统治地位、快速发展且迅速扩张之路，全球化的商业巨头伴随着高成长性和全新的商业模式，办公空间均需满足变革中的新需求。

以上新兴的“独角兽”企业和传统的全球型跨国企业为适应其办公方式的革新，同时亟需新的形象代言、企业精神与文化载体、全球运营和管控核心。新的总部园区建设势在必行，于是近年来掀起了建设全球总部办公园区或者区域总部办公园区的热潮。

GE中国近年来同样在北京经济技术开发区建设了新的“GE北京科技园”，并于2016年10月13日举行了开园仪式。GE北京科技园项目对互联网时代新型办公模式进行了有益的探索，园区的智能化、社区化实践有一定的代表意义，本文将对GE北京科技园的设计理念和设计建造过程进行分析总结，并对近年建设的类似总部进行比较，对未来的办公和园区型企业总部的设计进行探讨，希望可以为其他项目设计提供借鉴。

二、“筑梦之城”——GE北京科技园设计

世界五百强美国GE通用电气公司，1878年由爱迪生创办，目前是全球最大的跨行业经营的科技、制造和服务型企业之一。2015年销售收入1483亿美元，世界五百强排名第24位。GE中国在北京经济技术开发区打造其中国北方区的运营和研发总部——GE北京科技园（GE Beijing Technology Park，BTP），使得之前分散于北京各个办公地点的包括航空、发电、医疗、油气、可再生能源、运输、能源互联、研发及数字等不同的业务部门能够聚集在同一个屋檐下，同时GE中国将GE医疗集团全球的研发设计、中国区的运营管理及部分销售统一整合在一起置于新建设的GE北京科技园中。GE北京科技园于2016年10月正式落成启用。

本项目位于北京经济技术开发区65街区65M6地块，北侧临荣昌东街，东侧临同济南路，西侧临65号支路；南北长约284米，东西长约177.2米，总用地面积约50113.1平方米。用地规整方正，地势基本平整。用地周边以工厂及办公建筑为主，景观环境以市政绿化为主。本项目总建筑面积74200平方米，其中地上58400平方米，地下15800平方米，容积率1.17，建筑密度29%，绿地率15%。建筑主要功能：实验、研发、运营。建筑层数：地上5层，地下1层。建筑高度：28.8米（室外地面至檐口）。结构形式：钢筋混凝土框架剪力墙结构。

GE北京科技园项目由坤鼎投资管理集团股份有限公司作为投资方承担项目整体建设，通过建筑设计竞赛于2013年选定由北京建院约翰马丁国际建筑设计有限公司承担项目的全程设计服务。项目于2014年4月开始建设，2014年9月主体结构封顶，2015年4月完成外幕墙，2016年6月完成精装修并开始运营调试，2016年9月整体移交，2016年10月园区正式启用。

本项目作为GE医疗在北京乃至整个大中华区开展运营与研发工作的中心，有着十分重要的功能定位，同时也应当具有一定的示范和象征作用。

（一）筑梦之城——设计理念契合企业价值

梦想启动未来——GE企业文化魅力。本项目设计目标与GE的企业理念相契合，即遵循理性创新、面向未来、绿色生态的设计原则，为GE设计出一幢符合全球化企业形象的创新型总部、一处富有归属感的企业园区、一个孕育梦想的场所——一座“筑梦之城”。

图 1-2 方案鸟瞰效果图

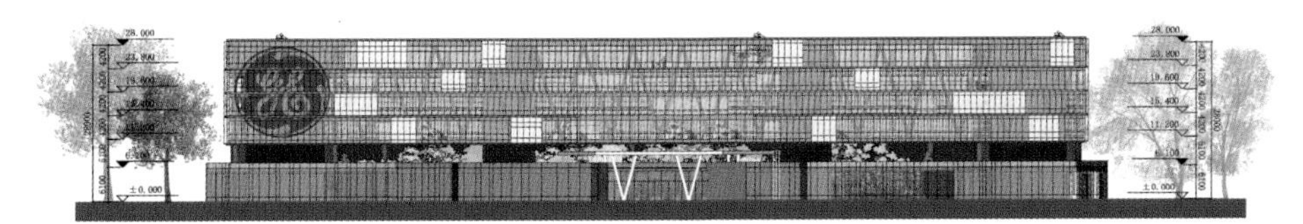

图 1-3 北立面图

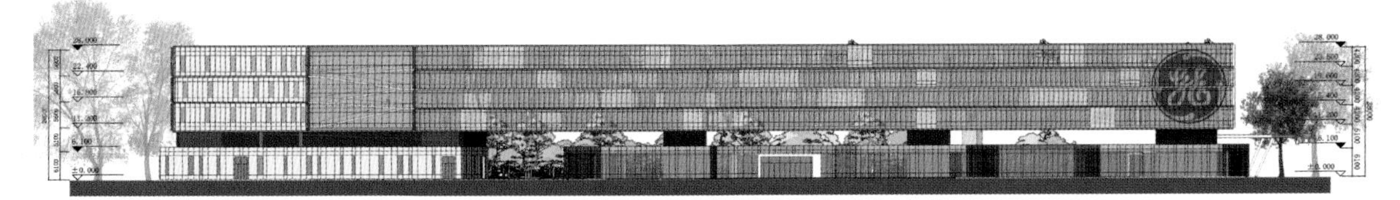

图 1-4 东立面图

图 1-5 北立面实景

图 1-6 北立面夜景

Wang Xiangdong

图 1-7 方案东北侧透视效果图

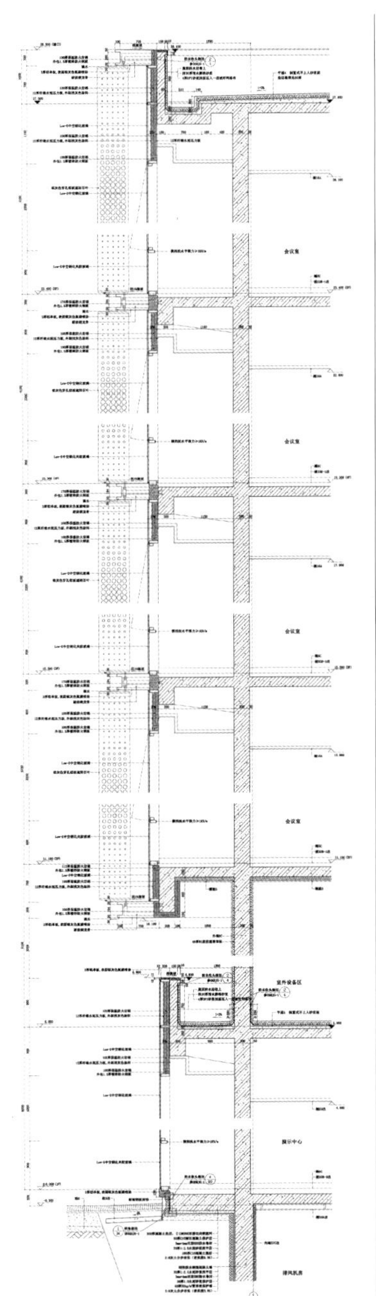

图 1-8 外墙构造详图

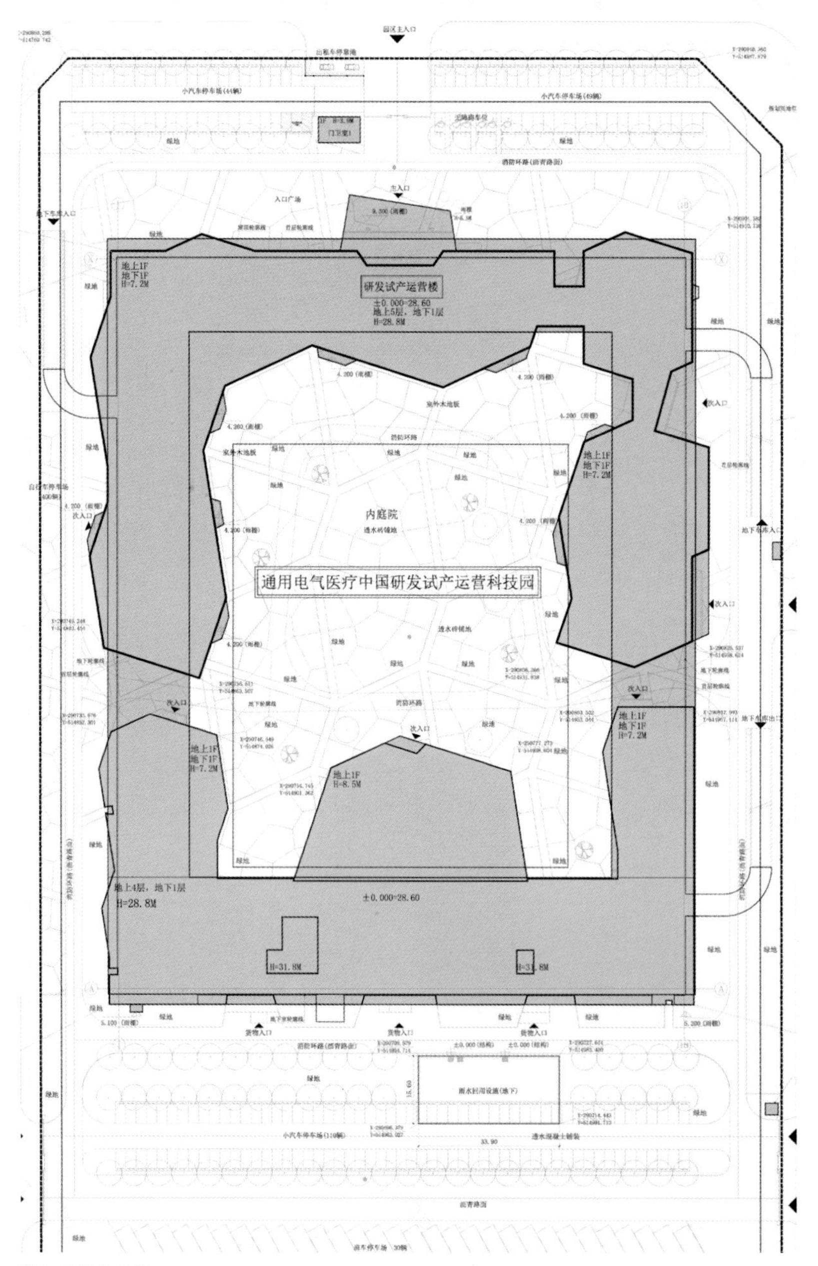

图 1-9 总平面图

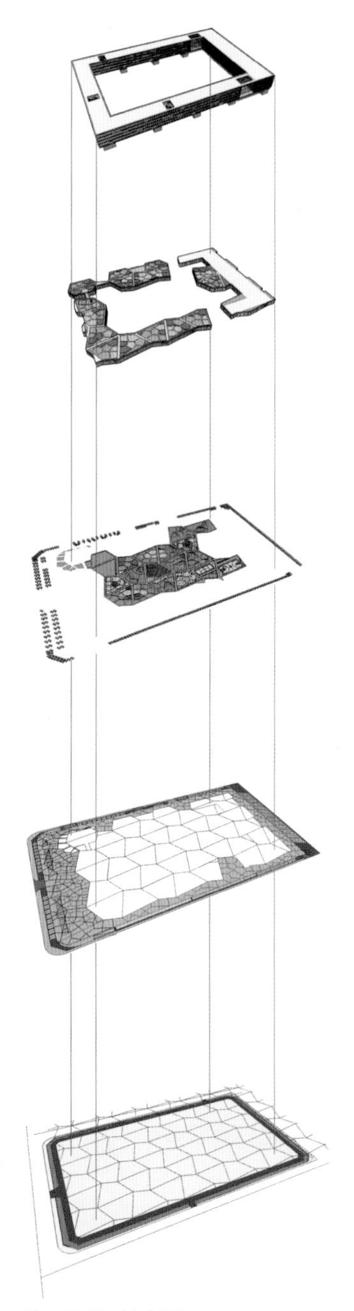

图 1-10 轴测分析图

空间是建筑的主体，也是建筑“物质功能”和“精神功能”的主要载体，无论何种功能的办公建筑，不仅要有相对应的文化定位，还离不开已经树立的标志性。标志性建筑不仅取决于社会地位，更重要的在于是否具有比建筑形式更重要的建筑精神。

建筑采用了中国传统的方形，是国际企业与中国传统“天圆地方”文化的集合，建筑营造的空中漂浮的合院，与北京的城市肌理和城市建筑文化有机结合，是美国企业文化与中国传统文化的完美结合。建筑形象简洁大气，与GE国际化、全球化的视野不谋而合。

建筑将大大小小的功能整合起来，以一个空中合院形式出现。以北京为代表的四合院的围合有多种文化精神及现实的原因，然而它的围合并不是封闭，庭院扩展了内部空间，使房屋摆脱了围护结构的制约。内与外、开放与围合、封闭与通透融合成为一个有活力的有机体，这是一个动态的对比并且它们彼此呼应着。围合布局使得建筑对外有完整、简洁的建筑形象，与城市交界面代表了一个企业的整体面貌，简洁、清晰、大气，更有利于标志性的塑造。

通过整合园区内外部用地和围合庭院，营造出属于园区自身的集中公园式景观，建造出具有高品质的室内外交融的复合空间。独到的布局使建筑不仅产生了外刚内柔的形象，同时也给使用者提供了更加近

图 1-11 西北侧夜景

08:00 员工乘坐班车来到公司

08:30 在咖啡厅用早餐

09:45 在办公室整理资料

10:20 汇报研究项目的进展

12:10 和同事一起用餐

13:00 在景观优美的内院散步

14:00 准备下午交流

15:00 在舒适的环境下办公

16:30 在体育设施上锻炼

18:00 乘坐班车离开公司

图 1-12 GE 人的一天（组图）

人和丰富的空间感受。建筑形象要服从建筑的功能布局，空间合理的穿插与叠合不仅可以塑造出特殊的“景观”及现代办公形象，还提升了高雅企业的文化品质。

建筑首层轮廓与“生命—脉络—细胞”理念相契合，功能空间与景观空间交互共生，围合成集中式的中心景观庭院，并形成有机的形体。建筑整体尺寸为174.03米×139.3米，首层设置可穿越内庭院的通道。10个垂直核心筒作为2至5层的支撑结构，联系首层和上面的研发试生产运营中心，形成夹层及室外设备区。2至5层建筑形体严整、方正，整体造型以方形为主。办公研发区2至4层为模块化办公空间，每个模块均设置有灵活开敞的办公区、即兴会议室、视频会议室、电话亭、合作区、储物柜、打印室、自助咖啡和茶水间。

办公研发区首层外立面采用高反射玻璃幕墙突出建筑的开放性，一方面可以更好地与周边环境相融合；另一方面可以强化空间理念，营造出空间无界、创新无限的GE绿色办公环境。实验区外立面以灰色质感涂料墙面为主，设置规则的条形窗。整个建筑表面以均匀的竖向立面肌理为主，建筑立面完整统一、优雅简洁、端庄大气。办公研发区2至5层外立面为简洁的隐框玻璃幕墙及色彩渐变的穿孔铝板遮阳百叶，既有利于遮阳，同时在一定程度上也降低了室外噪声对办公环境的影响。面向内庭院主要以条形隐框玻璃窗和灰色质感涂料墙面为主，以局部挑出的玻璃盒子将长的形体打碎，活跃了建筑立面形象，给人以深刻感受。

图 1-13 东北侧实景

大巴流线
大巴流线
大巴停车场

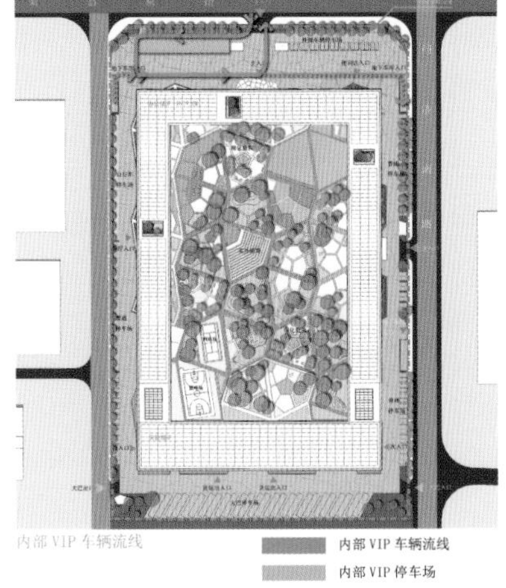

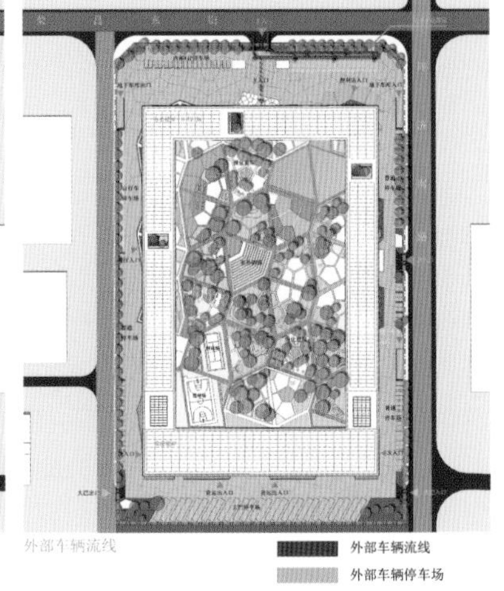

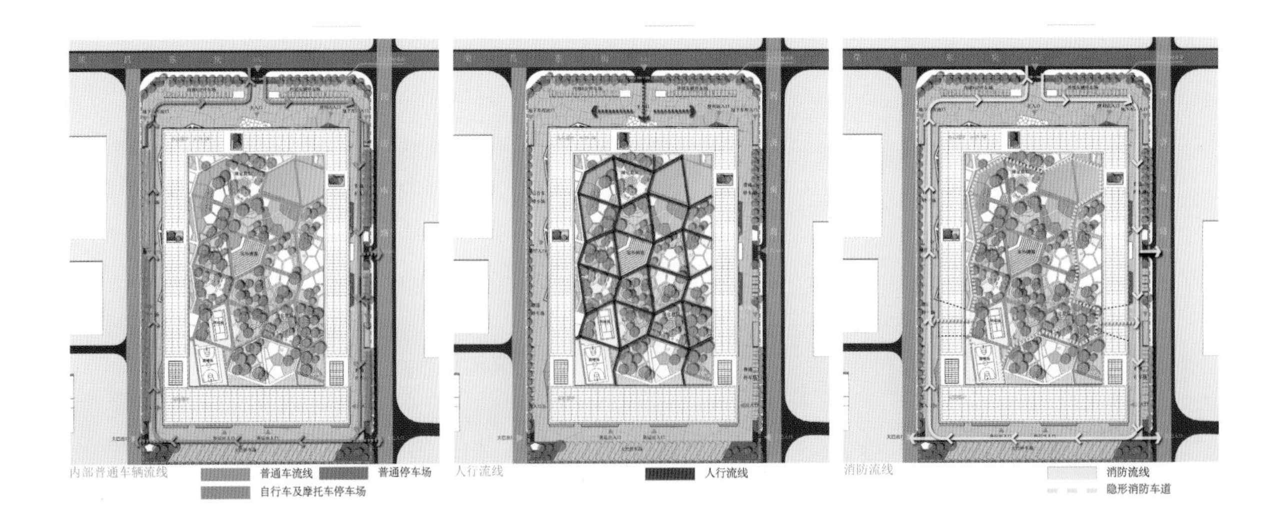

图 1-14 流线分析（组图）

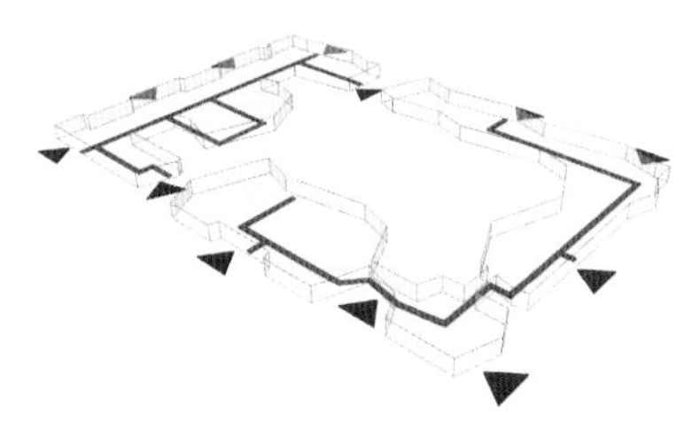

出入口——主要出入口的合理设置有利于更好地组织交通流线，使建筑内外有序地交流，形成合理统一的建筑空间

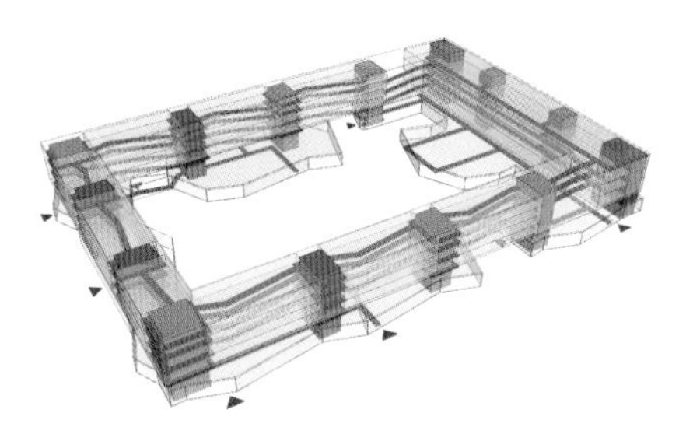

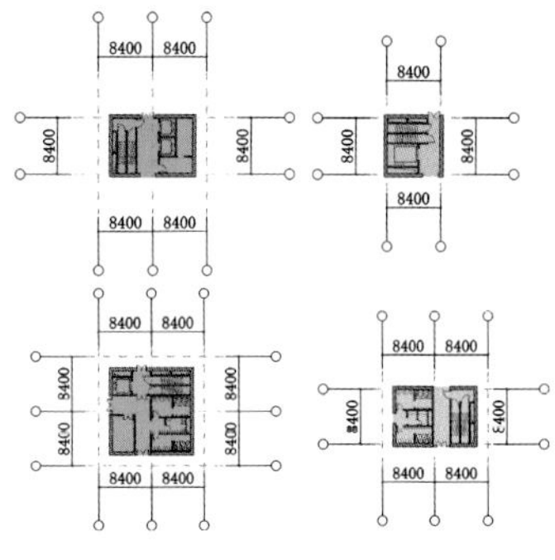

主交通核
辅助交通核
实验室货运交通核
主要出入口
交通流线

竖向交通——根据空间的开放与私密和功能的需求，竖向交通核有主有次，有利于合理地利用公共资源，达到绿色有机的立体交通体系

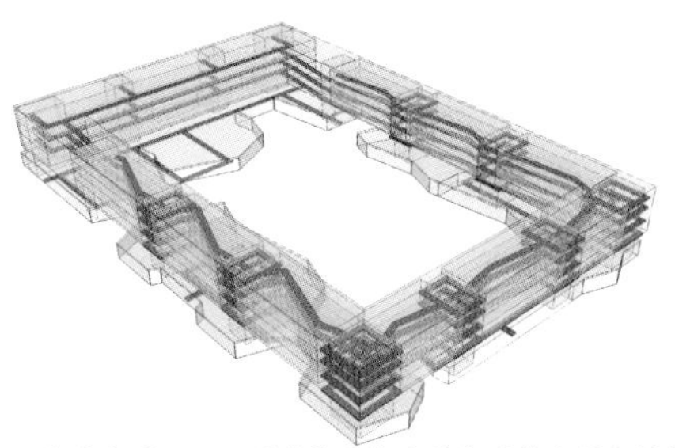

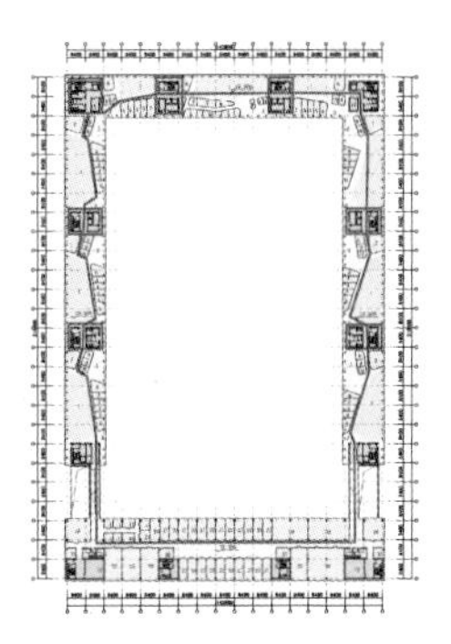

水平交通——环形的交通体系使各功能分区之间相互连接，办公区和实验区通过一系列的开敞和交往空间相连接，缩短了科研实验与办公空间的距离

图 1-15 交通组织分析（组图）

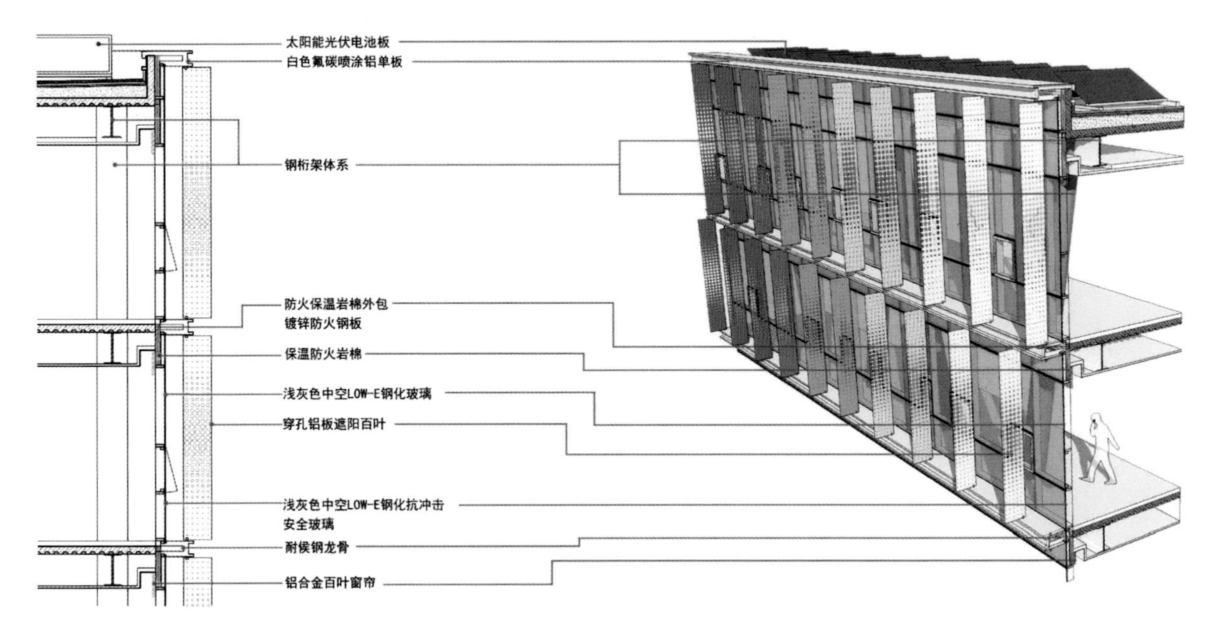
太阳能光伏电池板
白色氟碳喷涂铝单板
钢桁架体系
防火保温岩棉外包
镀锌防火钢板
保温防火岩棉
浅灰色中空LOW-E钢化玻璃
穿孔铝板遮阳百叶
浅灰色中空LOW-E钢化抗冲击
安全玻璃
耐候钢龙骨
铝合金百叶窗帘

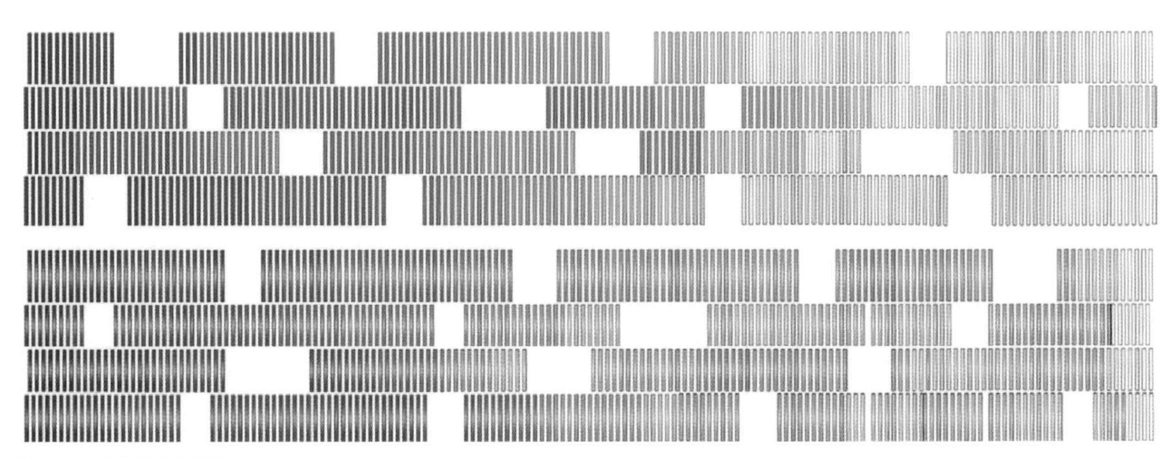

图 1-16 立面分析（组图）

图 1-17 立面构造细部

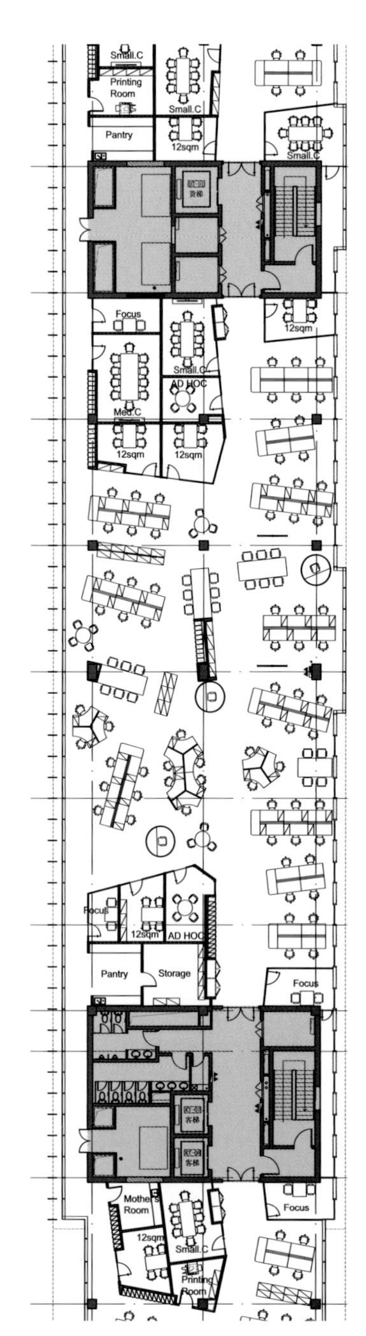

图 1-18 办公单元平面放大图

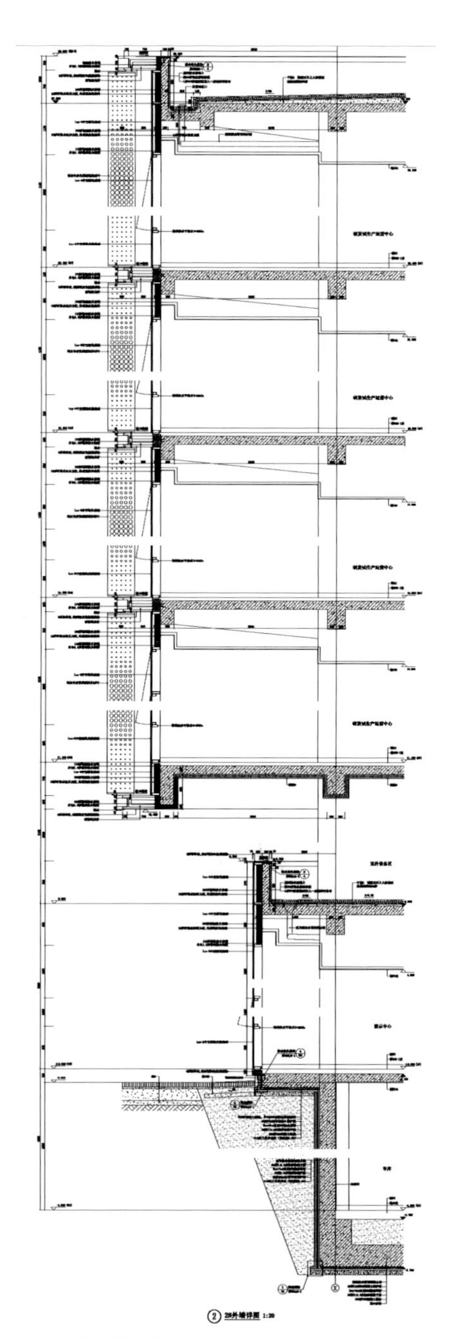

图 1-19 外墙构造详图

图 1-20 西南侧入口实景

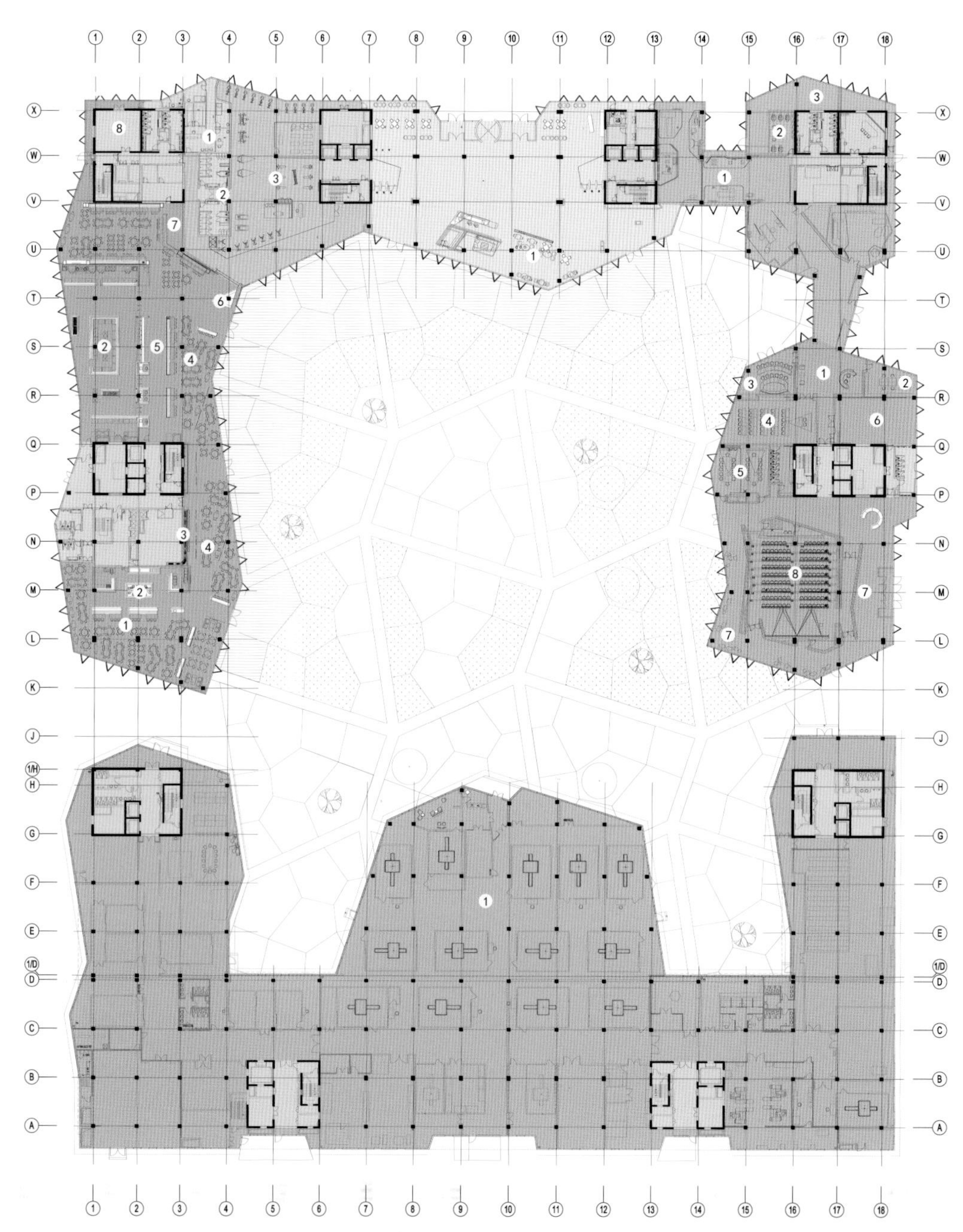
1 2 3 4 5 6 7 8 9 10 11 12 13 14 15 16 17 18
X W V U T S R Q P N M L K J 1/H H G F E 1/D D C B A

图 1-21 一层平面图

（二）平等、交流、共享、可持续的办公空间营造

GE大中华区总裁兼首席执行官段小缨女士对于理想的办公空间设置有特别的思考。除了在GE北京科技园构建新型的平等的办公空间之外，还要考虑公司未来的快速发展对办公空间的迅速改变有大量的弹性要求。

平等管理就是在企业(组织)管理中人人处于平等地位，尊重他们，不要把他们视为管理或监督对象，充分把他们的潜能开发出来，为企业(组织)目标服务。GE北京科技园从物理空间上对平等与交流管理组织进行了尝试。

GE北京科技园办公空间标准单元取消了全部的隔墙和格子间，工位布置有别于传统一字形、矩形布置，而是采用比较活泼、自由的组合方式进行异形排布，看似随意分布着的由一个个办公桌和工作空间组成的“岛屿”，与开敞式办公区吧台式合作空间和休息空间以及白色的云形格栅吊顶相互呼应，使整个办公空间的氛围更加轻块、自由、舒适、温馨。整个园区2至4层办公空间区域均不设置管理人员办公室；经理层和员工根据自己的工作任务和工作行为，按预约时间选定自己的工位；5层设置总裁、高管层行政办公区，取消总裁、高管层独立办公室，提高效率，创造

图 1-22 主入口实景

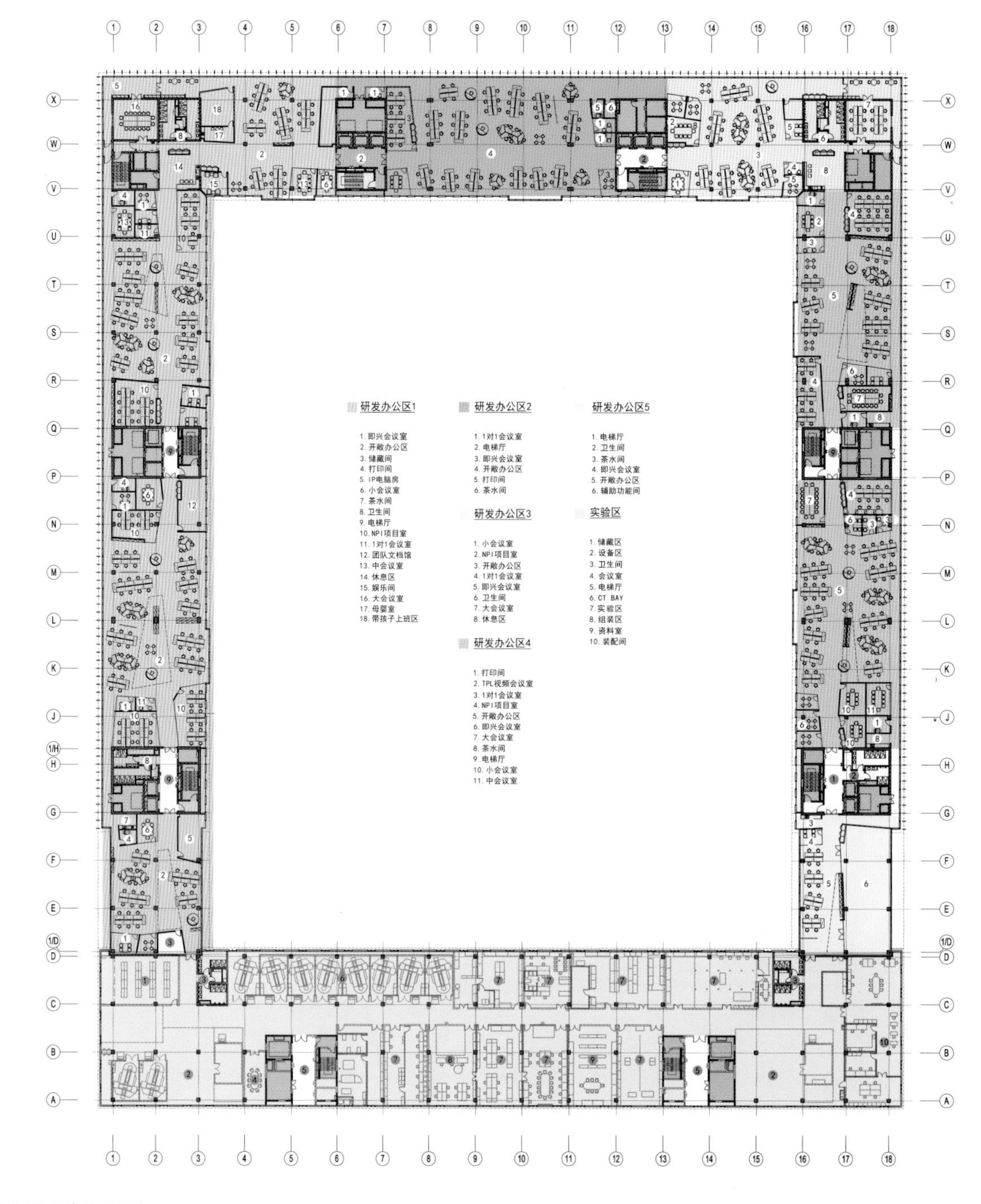
研发办公区1
1. 即兴会议室
2. 开敞办公区
3. 储藏间
4. 打印间
5. IP电脑房
6. 小会议室
7. 茶水间
8. 卫生间
9. 电梯厅
10. NPI项目室
11. 1对1会议室
12. 团队文档馆
13. 中会议室
14. 休息区
15. 娱乐间
16. 大会议室
17. 母婴室
18. 带孩子上班区
研发办公区2
1. 1对1会议室
2. 电梯厅
3. 即兴会议室
4. 开敞办公区
5. 打印间
6. 茶水间
研发办公区3
1. 小会议室
2. NPI项目室
3. 开敞办公区
4. 1对1会议室
5. 即兴会议室
6. 卫生间
7. 大会议室
8. 休息区
研发办公区4
1. 打印间
2. TPL视频会议室
3. 1对1会议室
4. NPI项目室
5. 开敞办公区
6. 即兴会议室
7. 大会议室
8. 茶水间
9. 电梯厅
10. 小会议室
11. 中会议室
研发办公区5
1. 电梯厅
2. 卫生间
3. 茶水间
4. 即兴会议室
5. 开敞办公区
6. 辅助功能间
实验区
1. 储藏区
2. 设备区
3. 卫生间
4. 会议室
5. 电梯厅
6. CT BAY
7. 实验区
8. 组装区
9. 资料室
10. 装配间

图 1-23 标准层平面图

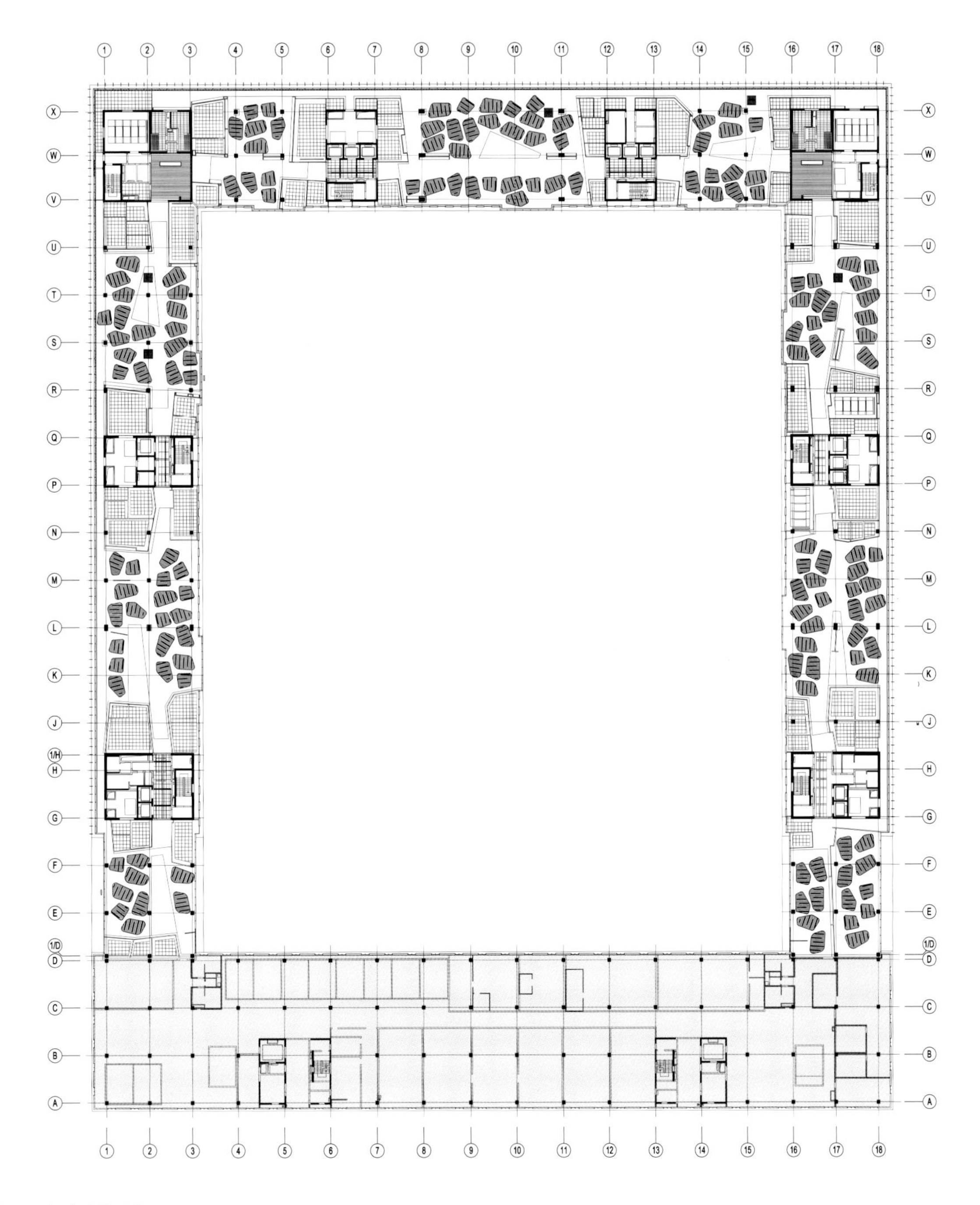
1 2 3 4 5 6 7 8 9 10 11 12 13 14 15 16 17 18
X W V U T S R Q P N M L K J 1/H H G F E 1/D D C B A

图 1-24 标准层吊顶图

更多交流空间。

在这种办公空间里，员工们不用再整齐划一地坐成数排，大家可以闲聊、走动，营造出一种随意的气氛。此外，这种布局也同样传递着自由民主的气息。获得晋升的员工或许还会留在原地或在附近位置办公，避免了因搬进独立办公室而引起的嫉妒与不便。GE北京科技园通过这种物理空间的设计，减少了管理层和员工的隔离感，增加了更多、更自由的交流方式，从而保证企业能够以最快的速度将决策传达至员工，进而提高企业效率，并建立起富有弹性的新型管理模式。

同时，办公空间取消了所有的部门界限，弹性的办公空间使各个部门能共享公共设施，在有需要的时候各个项目团队可以坐在一起讨论，而不是被格子间和部门界限所束缚，为空间的共享和可变创造足够的条件，部门的规模扩展和压缩不需要进行建筑的物理改变。每个标准层还设置有5个大跨度无柱办公空间，分别为20米×24米1个、17米×21米1个、18米×24米1个、16米×17米2个，此类大跨度无柱空间也为办公布局提供了最大的灵活可变性。

办公空间标准单元均应用网络地板，网络地板安装方便、自然形成布线槽，同时比传统高架地板更节

图 1-25 可变的办公空间（组图）

省净空，布线更灵活。网络地板在办公空间的应用，更便于灵活调配固定与移动工位，需要时能即时更新、做出改变，设计灵活的办公空间为未来多变空间创造了无限拓展的可能。

内部会议室为共享设计，取消了部门独立会议室，根据工作任务、参会人员数量和时间，对会议室进行预约。会议室及会议资源可以得到更好地共享协调，不仅节省了会议组织者的时间，提高了会议空间的使用效率，而且提升了企业的形象。

从实施效果来看，设计实现了GE内部平等的组织关系；交流空间的营造为未来的办公空间的快速改变营造了足够弹性；共享公共设施达到了节约办公空间的目的。

图 1-26 员工餐厅（组图）

（三）办公园区的生态圈营造

1.社区营造

GE北京科技园一层主要承载着接待大堂、企业文化和产品展厅、设备齐全的健身房、丰富多样的餐厅、星巴克、不同类型的会议室、医务室、便利店、托儿室、多功能厅、培训室等功能。大堂区以“过去、现在、未来”的概念打造了3个从吊顶上悬挑下来的全息屏幕，展示着GE在全球和中国的相关影像以及对未来发展方向的展望；展示区呈现的是GE全球最新产品和互动技术，展示区提升着企业的文化品质，也是企业彰显科技统治地位的纪念馆。

首层大型报告厅、会议室、培训室采用集中布置的方式，为办公园区多元化的活动和交流提供可能。宽敞整洁的员工餐厅，体现出GE对员工的关怀，种类丰富的餐饮选择和空间带来清新舒适的就餐环境，使员工更加积极地投入工作。引进“星巴克”入驻北京GE科技园，营造出一种温暖而有归属感的文化，欣然接纳和欢迎每一个人。坦诚相见，互尊互敬，让员工相互喜爱并分享自己的新发现，激发并孕育人文精神。

设置功能齐全的健身房，缓解工作压力，让员工在工作之余能更好地放空和锻炼自己。运动能让人感觉状态更好，精力更充沛，通过运动，可以提升自信心、工作效率和记忆力。首层便利店设计，解决员工即时性、小容量、急需性消费等问题。首层设置自助证明打印机，节约人力资源成本，员工自助刷卡可高效打印带有公司人事章的收入证明和在职证明。儿童活动室和医务室保障员工应急需求，儿童活动室的设置解决了寒暑假期间员工子女无人照顾等问题，医务室的设置解决了员工治疗小病小伤的需求。

每个办公空间单元均设置有小型自助型咖啡机、茶水区和开放式吧台，员工刷卡即能享用即时咖啡或茶水；电话亭的设置解决了员工对外沟通的私密性和对周围员工的影响问题；安静位置都配备

图 1-27 医疗设备展示空间

患者安全
Patient Safety
120 db

了带有专属门禁的母婴室，室内采用环保吸音的毛毡进行墙面装饰。办公园区中央绿化庭院设置步行漫道和栈道，员工能在绿树成荫的园区里工作，容易找到灵感，实现了在运动中享受生活和工作，在生活工作中享受运动。

GE北京科技园围绕工作、生活，会聚各部门的服务和生活配套设施中心，提供工作、就餐、休息、培训、娱乐、健身、医疗、展示诸多空间等设施，同时通过多元化的活动、交流，促进并形成内部办公生活群。所以，GE北京科技园不仅是一个物理的办公空间，更是一种全新的总部办公生活方式，塑造了一个总部办公生态社区。

2.园区景观设计

景观与普遍意义上的物质与能量资源相比，是一种具有较高层次的信息资源。景观资源代表的范围十分宽泛，它必须经过艺术创作和科技加工，才能更具观赏、文化或科学价值，同时生态景观与人文景观都要服从于城市建筑的总体设计思想。通过整合外部环境用地，围合出内部庭院，以此营造出属于园区自身的集中的高品质公园式中心景观，为使用者提供更加近人和丰富的空间感受。外侧是以硬质铺装为主的广场景观，内侧为集中式的中央园林景观。景观设计的

图 1-28 健身房

图 1-29 星巴克咖啡厅

图 1-30 一层大堂

图 1-31 同传会议室

概念是挖掘生命、脉络及细胞的内涵与联系，在合理的尺度及结构要求下，纹理在基地内全面铺开，贯穿于中央园林和外部场地，景观形态与首层建筑的外部形态有机结合，形成一种丰富活跃的、富有生命力的空间效果，实现景观共享。

（四）智能互联的社区运营

在新的互联网时代下，GE北京科技园的智能化设计和互联互通的理念贯穿整个设计和园区运营管理中。园区的智能化系统主要包含以下四个部分。

①智能安全管理。园区对工作人员采用分级的身份认证系统，不同工作人员有不同的认证权限。进入科技园区的非工作人员需通过预约系统进行来访者与被访者的智能登记，打印二维码身份认证卡，预约系统自动传输至被访者，来访者持有二维码身份认证卡进入园区，其权限仅能进入相应预约区域。

②智能工位预定系统。引入智能工位管理系统，预定便捷，多种模式，实现可视化工位地图设计和管理。可通过地图查找工位推荐，通过邮件、微信、移动应用程序、电脑网络、一体机多终端预定；规范管理，优化资源，实行工位签到、延时和注销，为来访者预定，临时搭建团队工作区；提高使用率，降低费用，灵

图 1-32 医疗展示序厅

图 1-33 远程医疗展示厅

图 1-34 视频会议室

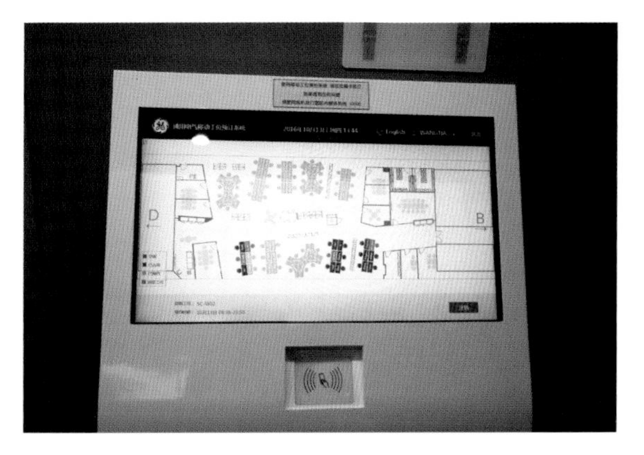

图 1-35 智能选位机

活调配固定与移动工位，部门专属或共享工位资源等。

③智能会议预定系统。多媒体会议室预约管理系统，可以让会议组织者不受时间、地点的限制，轻松地通过网络在后端实现会议室的查询、预定、管理、使用等。

④智能楼宇控制。园区内设有完善的楼宇自动化管理系统，空调自控系统为BA系统的主要组成部分。通过BA系统的数据通信，可以对冷热源、空调系统的各种参数进行监测和设定，并对所有的空调通风设备进行监视和故障报警，还可以通过与其他系统的通信，确定人员使用情况，结合室外的气象参数，综合确定系统工作模式。

园区运营引入国际综合设施管理服务公司，依托多年在国际市场所积淀的专业知识和经验、高质量的管理和服务品质、国际化的人员储备以及对跨国企业和国际品牌的理解和良好的合作关系，对园区进行全程的综合设施管理服务。国际化的综合设施服务除传统项目外，还在保障园区设施运行和安全以及为园区员工工作和生活质量提升创造条件。

（五）GE医疗实验室设计

在整个园区南侧呈一字形布置的是相对独立的GE实验楼原材料及成品库房和试生产中心，主要是对核磁共振仪、CT机、X光机等新研发的产品进行组装、调试、测试及样品的试生产以及对关键零部件进行测试。首层设置有接待室、洽谈室、16个核磁共振仪Marketing Bay、扫描架组装区、介入治疗实验用房及各种配套设备机房；2层主要设置为10个CT Marketing Bay、CT电气实验室、测试区及各类设备机房；3层为33个CT Marketing Bay、CT储藏区及各类设备机房；4层为27个X射线Marketing Bay、X射线ME实验室、X射线绘图室及各类设备机房。

设计团队和GE医疗研发团队进行了紧密的协作，对于专项实验室的各项特殊要求均提出了详尽的解决方案，整个实验室的设计均严格按照实验的特殊工艺要求和结构荷载、空调、散热、供水、供电、污染、防辐射等要求完成。该实验室是GE医疗全球最重要的研发实验室之一。GE中国医疗研发中心1000多名工程师组成的GE医疗技术研发团队，正为中国和全世界开发领先的医疗产品与技术。

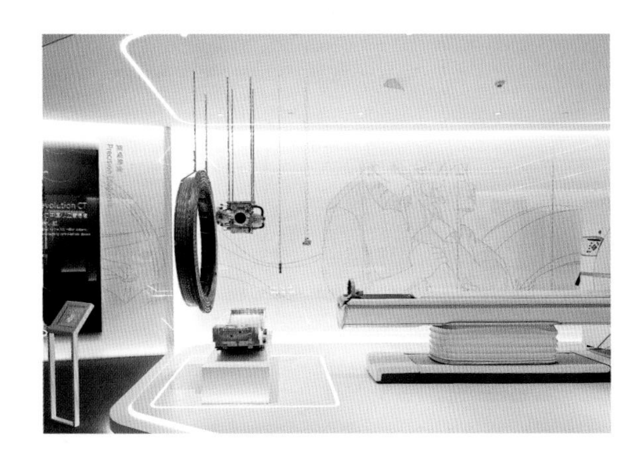

图 1-36 医疗设备展厅（组图）

（六）结构设计

结构体系采用钢筋混凝土框架—剪力墙结构，局部大堂及报告厅等位置因建筑功能需求，有24.3米的大跨度，采用型钢混凝土梁柱构件，降低梁高以满足建筑要求。建筑整体呈口字形，结构在“±0米”以上设抗震缝脱开，以一个一字形单体与一个U字形单体围合成现有建筑形态。U字形单体部分平面上利用8个建筑交通盒形成的剪力墙筒体控制结构的整体位移。建筑外围尺寸为174.03米×139.3米，U字形及一字形结构单体单边长度均超过100米，采用加强框架梁腰筋、楼板配筋，设置后浇带及规定合拢温度等方式控制混凝土因温度、干缩等因素产生的裂缝。

1.基础

为解决沉降差，同时又能满足建筑功能的要求（主楼与实验楼荷载差异较大），本工程采用CFG复合地基—有梁筏板基础，采用调整CFG桩长、桩间距等方式调整地基刚度及承载力，同时在计算时模拟地基刚度，设计地基梁及筏板，以应对上部结构有所差异的地基反力，同时节约部分建设成本。

2.上部结构

因业主提出的楼面荷载较大，约10千帕，实验楼采用主次梁楼面体系以降低板跨，且需考虑天车等多种荷载工况，均单独验算复核设计。

（七）给排水系统

生活给水水源由1路DN150市政给水，采用生活调贮水箱加变频调速泵组联合供水。中水由DN100市政中水提供，用于冲厕、绿化浇灌。集中生活热水采用太阳能，辅助热源为城市热力。本工程室外生活污水和生活废水为合流排水系统，园区内设集中污水处理站。屋面设虹吸雨水排水口。园区设2处雨水调蓄池，1处调蓄入渗池，1处调蓄回用池。

地下1层设消防专用贮水池。水池贮存消防用水量1150立方米。室外消防用水量为40升/秒，消防泵房内设置2台室外消防供水泵及一套增压稳压装置。室内消防用水量为30升/秒，采用贮水池、消防泵、高位水箱联合供水方式。自动喷淋系统用水量为100升/秒（符合FM认证的需要），研发区采用标准K80喷头，实验区和地下均采用K115喷头，作用面积根据FM认证的要求确定。

（八）空调系统

制冷机房设于地下1层。采用3台制冷量为1758千瓦的离心式冷水机组，1台制冷量为673千瓦的螺杆式冷水机组（仅值班使用），供应6/13摄氏度空调冷水；其冷源侧为定流量运行；负荷侧设二级泵，采用变频调速泵组使系统变流量运行。实验区内的密闭区域采用独立的多联机空调作为冷源，全年供冷。热力站设于地下1层。由城市热网引入1根DN150蒸汽管道作为一次热源，市政蒸汽压力为0.8千帕，减压后供楼内使用。通过3台汽水热交换器和热水循环泵，为建筑物供应供回水温度为60和45摄氏度热水。热水循环泵采用变速调速控制方式。

表1-1 不同功能区的空调形式

功能区名称	供暖空调系统
厨房	排风＋补风（带加热盘管）＋散热器
研发运营室、会议室	外区风机盘管系统，内区变风量全空气系统（过渡季节可全新风运行）
大厅	变风量全空气系统（过渡季节可全新风运行），冬季设地板辐射值班采暖
展厅、多功能厅、餐厅	变风量全空气系统（过渡季节可全新风运行）
实验区（Bay外部分）	风机盘管加新风系统
实验区（Bay内部分）	多联机空调加新风系统（全年供冷，24小时）
电梯机房、电话控制室、消防控制室	独立冷源的空调器
IT机房	机房专用恒温恒湿空调

办公区域为倒U形区域，除个别会议室外，均为开敞式办公区。标准层每层分为9个独立的空调区域。空

图 1-37 内院交流空间

调外区采用四管制风机盘管系统，内区采用单风道变风量末端的全空气系统。根据负担区域负荷的差异，每台空调机组风量为8000~13000立方米/小时。每个空调区域内的单风道变风量末端在10个以内，单个变风量末端的最大风量为1300立方米/小时。新风集中由屋顶引入室内，冬夏季最小新风工况时，新风集中由屋顶热回收机组处理后，送至各层空调机组；全新风工况时，新风不经处理，直接由空调机组送至室内。

由于每个空调分区内的变风量末端数量均在10个以内，数量较少，使得楼宇自控系统与每个变风量末端通信联网成为可能。因此，本项目选择总风量法作为变风量系统的基本控制方案。考虑到总风量法的通信量较大，在通信故障或者系统有其他异常时，辅助采用定静压控制法。所有风机盘管均要求纳入BA系统管理，开敞办公区的风机盘管采用分组群控。

1.自然通风模式

在室外温度为10~25摄氏度，且污染物浓度为轻度污染以下时（空气污染指数API≤150）采用，为全年推荐的运行模式。

本项目有非常适合自然通风的客观条件。

①由于地块要求的容积率较低，导致建筑的体形系数较大。这样建筑的布置相对松散，非常有利于自然通风。当然，这也有不利的一面，本项目体形系数大，且采用了玻璃幕墙，导致建筑通过围护结构的得热较多，在设计过程中，建筑师考虑了有效的外遮阳措施，在一定程度上弥补了不足。

②本项目为口字形的布局，开口位置的选择经过了精心的设计。首先，口字形布局有利于防止冬季建筑物两侧风压差过大，导致建筑漏风量增加；其次，本项目进入内庭院的通道位于建筑的东南和西南方向，过渡季东、南侧主导风向时，有利于气流进入内庭院，增加建筑物两侧风压，进而达到有效的自然通风效果。

③本项目采用的“穿堂风”模式是效率最高的通风方式。70%以上的区域为开敞办公区，且标准层平面最大进深为20米，层高为4.2米，满足“穿堂风”通风模式下进深不大于层高5倍的基本要求。

2.机械通风模式

在非空调季，污染物浓度为轻度污染以上时（空气污染指数API＞150）采用。此模式下，所有空调机组均无冷热水供给，全新风运行，关闭热回收机组，并开启屋顶的集中排风机。

3.空调通风模式

空调通风模式有两种典型工况。

①最小新风工况，屋顶热回收机组开启，通过新风和排风系统的定风量阀控制室内的最小新风量，外区风机盘管系统和内区全空气变风量系统同时工作，且控制夏季外区温度为（26±2）摄氏度，内区温度为（24±2）摄氏度；冬季则控制外区温度为（18±2）摄氏度，内区温度为（20±2）摄氏度。

②由于人的活动很难被限制，在气候条件允许时，办公区内的员工总是倾向于开启外窗。这种情况

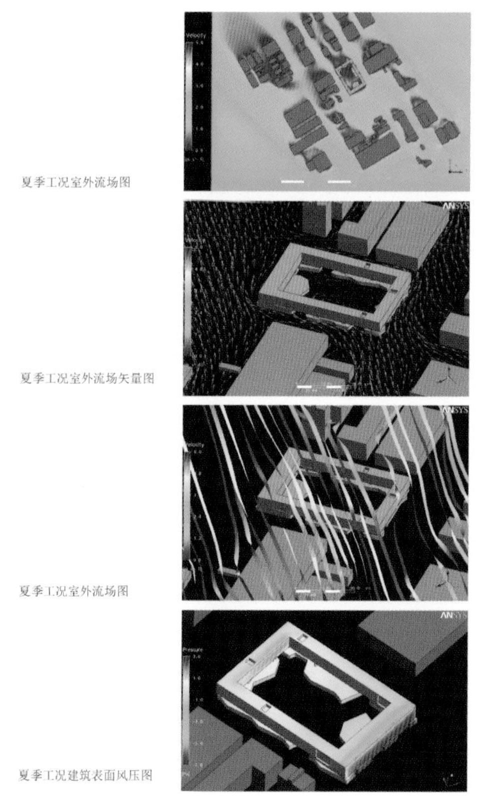

夏季工况室外流场图

夏季工况室外流场矢量图

夏季工况室外流场图

夏季工况建筑表面风压图

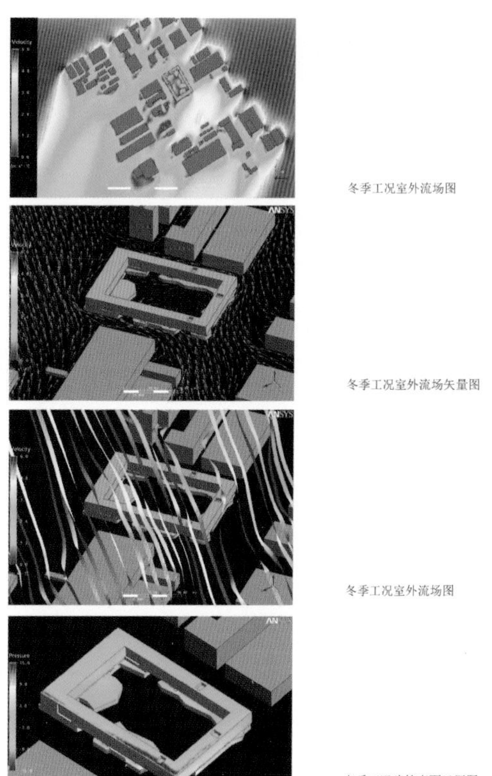

冬季工况室外流场图

冬季工况室外流场矢量图

冬季工况室外流场图

冬季工况建筑表面风压图

图 1-38 风环境模拟分析图

下，一般室外气候条件都比较适宜，可以忽略建筑的空调内外分区。该工况下，关闭外区的风机盘管系统和屋顶热回收系统，内区全空气系统可以适当增加新风量。

4.设计条件下的能耗预测与分析

本项目利用DOE-2.2 eQuest软件进行了建筑的全年能耗评估。围护结构参数、设备选型参数均依据设计施工图；人员、设备功率及部分照明密度等参数设定依据《公共建筑节能设计标准》GB 50189—2005设置；使用时间表的设定则依据ASHARE 90.1 2007用户手册设置。

计算得到本项目全年总耗电量为849.0×10^4千瓦时（对应的ASHARE基准参照建筑的总耗电量为1091.2×10^4千瓦时，相比基准建筑节能约22%），单位建筑面积能耗114千瓦时每平方米每年，除去采暖能耗，单位建筑面积能耗为95千瓦时每平方米每年。

（九）建筑电气

本工程建筑形式为口字形多层建筑，南北长174.03米，东西长139.3米，为解决供电半径过长、压降损失过大的问题，分别在南侧及北侧设置变电室。南侧变电室邻近冷冻机房，其中设置两台变压器做季节负荷专供冷冻机组用电，以解决与常用负荷合用变压器造成的冬季负荷低负载率运行的不经济运行方式。本工程中所有一级、二级负荷均引自两台不同的变压器，在为设备供电的配电箱（柜）中切换，消防负荷在末端切换，保证供电的可靠性。大容量电机采用降压启动，减小启动电流对配电系统的冲击影响，延长设备运行寿命，降低功耗。办公部分的会议室、职工餐厅、报告厅、大开间办公区、室外景观照明等处采用智能化照明控制系统。

1.建筑智能化

在4层设置综合布线总系统设备，全楼均设置内、外网，并实施物理分隔。主要由语音、数据、图文、图像等多媒体综合网络构成，使大楼办公自动化、通信自动化等统筹规划。建筑设备监控系统采用直接数字控制技术，对全楼的暖通空调系统、给排水系统、照明系统进行监控；对电梯系统及供电系统进行监视。

火灾自动报警系统的保护等级按一级设置，在南楼首层设置消防控制室，消防自动报警系统按两总线设计，设有消防联动控制及火灾应急事故广播功能。安防监视中心与消防控制室设在同一房间内，公共区域设置保安监视系统，采用用户多媒体图形软件控制一体化电视监控，含自动录像、录音等功能。

2.绿色节能、环保措施

变电室位于负荷中心。合理选择变压器容量，按经济电流密度合理选择铜芯导线截面以降低线路损耗。尽量减少负荷不平衡度，以减少变压器损耗。选择节能型电器设备：采用高效节能灯及其光源，照明采用智能照明控制系统；设置楼宇自动化管理系统，以确保所控制设备处在最佳运行状态，提高效率以达到节能目的。

（十）节能、环保、安全

本项目用能选用的电力、天然气、市政热力、柴油，符合当地的能源供应条件。

本项目通过美国LEED绿色建筑（Leadership in Energy and Environmental Design）认证,取得LEED认证的建筑物具有下列优点：①建筑物营运费用平均降低8%~9%；②建筑物价值平均增加7.5%；③建筑物平均节能24%~50%；④二氧化碳排放平均降低33%~39%；⑤用水量平均降低40%；⑥固体废弃物平均减量70%；⑦创造健康舒适的工作环境，员工生产力平均提升2%~16%；⑧提升业主企业形象，增加与国际大公司合作之机。

根据GE的全球资产的保险要求，GE在全球的资产均需由FM Global承保，FM Global仅承保“严格受控风险”(Highly Protected Risks, HPR)财产，故项目在设计和施工的全过程均接受FM Global工程风险评估。FM Global的商业理念相信损前预防比损后补救更为重要，并相信大多数财产损失是可以预防

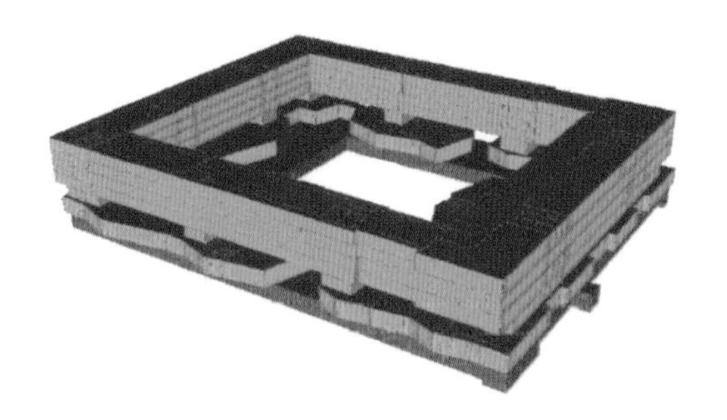

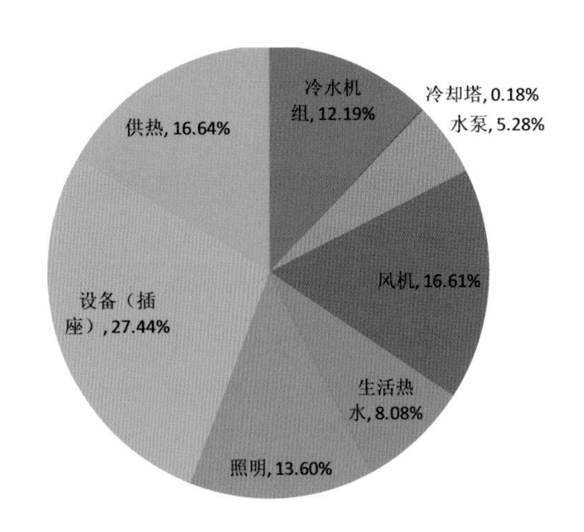

图 1-39 项目所获 LEED 证书及能耗模拟图

的，这一理念已成为公司经营的一项基本原则，所有运作都源于该理念。 FM Global以工程风险管理为主要手段发现、认识、评估、量化、控制并降低实际风险，评估的结果将在承保价格上得以充分体现。FM Global公司成立了一支精锐的队伍专门为所有FM Global的保险客户提供全面的风险评估工程服务，包括洪水、台风、飓风、地震等自然灾害和消防爆炸等人为的安全评估、设计图纸的全过程审核和施工现场查勘、工程产品的选用、根据FM Global技术规范和标准要求进行的消防和工程的验收服务等。FM Global会建议客户安装经过FM Global认证的工程产品，同时会对设计过程中涉及安全的环节进行严格的审查，对于FM Global的技术规范、技术标准与国内规范不一致的情况提出在遵守国内规范基础上的更安全的建议，设计图纸既要求满足国内的安全要求，特别是消防相关的规范要求，同时还要尽量满足FM Global的技术标准和规范，这对设计团队提出了更高的要求。实际上，项目在按照国内的验收标准验收之后，又经过了将近一年的时间才在2017年5月通过了FM Global的工程风险评估。

三、从GE北京科技园设计看大型企业总部与办公空间的未来

GE北京科技园为国际化的大型研发型企业自建研发总部园区打造了合理的内部共享办公模式，园区内配置全方位的服务和生活配套设施，成功打造了GE北京科技园内部办公生态社区；格子间的取消，设计灵活的智能数字化办公空间为未来多变空间创造了无限可能；网络地板应用，随时按需求更新，做出改变；取消管理者办公室，实现扁平化管理的理念，提高企业效率，创造更多交流机会和空间；智能运营和办公管理系统为企业对外高效沟通奠定基础。GE北京科技园作为目前国内最大规模的集团内共享办公项目之一，对传统跨国公司在新时代的未来办公模式进行

图 1-40 新浪集团总部办公楼

图 1-41 苹果公司总部办公楼

了有益的探索。

随着经济全球一体化的发展，企业的经营和生产早已超越了狭隘的地域局限，形成网状的发展结构。总部办公犹如这张网的节点，控制和引导着企业在周边区域的经营。在社会大变革的背景下，企业也在努力寻求转变，以获取更大的市场竞争力。企业总部建筑的功能，一方面是为了满足企业不断发展对办公空间的需求量增大及对办公空间环境质量提高的需求，另一方面是企业迫切需要一个平台来展示企业文化与产品。

在这种背景下，研发型总部建筑作为展现企业文化的载体，成为各大企业争相倾力打造的对象。总部办公建筑作为一个独立的类型，在这个时期逐渐发展成熟起来。本文所讲的总部办公建筑，是指企业或联合体作为业主兼使用者，在某一地域范围内最高级别的办公基地，通常体现为独栋或一组建筑，有其相对独立的办公环境。由于其主体单一，建筑设计过程中能够与最终使用者直接沟通，因此总部办公建筑与企业文化、核心价值观乃至整个社会价值观的紧密联系，也是本文关注的焦点。

根据GE北京科技园的设计总结和对新时代大型跨国企业的发展需求和总部建设的研究，我们认为研发型园区总部设计有它自身的特点，具体要凸显如下几个方面。

（一）标志性和精神核心

由于身兼“为企业代言”的作用，总部办公建筑总是如八仙过海般竭尽其能地展现企业自身的实力和个性。打造一个令人记忆深刻的、有传播效应的建筑形象无疑是总部办公设计的重要方式，无论是苹果公司总部的圆环，还是GE北京科技园长方形的“空中之城”、凤凰卫视的莫比乌斯、新浪的“∞”，都用简洁明快、可识别的建筑形象为企业代言。新型企业总部建设都会通过特有的方式和可识别性进行展示。

从新型办公空间的规划与设计研发出发，总部办公空间与一般单纯的新办公建筑不同，除了追求生态、智能、高科技品质外，更要强化自然、人文、文化的“标志性”品质，要创造出令员工自豪的建筑。

（二）企业文化的重要载体

美国学者伦斯·米勒在《美国文化精神》中说道：“企业唯有发展出一种文化，这种文化能够在激励中获得成功的一切行为，这样公司才能在竞争中获得成功。”这就是企业文化。

我们可以看到，苹果在对待追求产品细节上的极致同样保留到了苹果的新飞船总部大楼（Apple Campus 2）的设计上。苹果把Apple Campus 2 看

图 1-42 立面及企业 LOGO 细节

图 1-43 主入口侧视图

图 1-44 西北侧实景

图 1-45 内院实景

作一台巨大的iPhone——从管道、玻璃到电梯的按钮，都遵照了硬件设计那样的严苛标准。苹果新总部园区是苹果公司创新与冒险、注重细节和团队取向的组织文化的重要载体和集中展示。同样，GE北京科技园是GE“梦想启动未来”文化的承载，设计目标遵循理性创新、面向未来、绿色生态的设计原则，为GE创造了一处富有归属感的企业园区、一个孕育梦想的场所。园区的建设除了标志性之外，更重要的是如何体现公司的企业文化。

（三）创新的源泉——平等、交流的办公空间

美国麻省理工大学的研究表明：80%的创新来源于非正式的交流。乔布斯在其传记中说：“建筑空间也在说话，如果一个建筑物不鼓励交流合作，那在建筑里的公司就不能感受到由高‘碰撞率’引发的魔力，从而失去很多的创新。”

传统公司的员工，只在一个垂直部门相互交流，跨部门的事就不是自己的事了。现代的公司跨部门相互交流才能让公司的效率更高，创新力也能得到增强。如今，企业强调跨界交流，随时学习新领域的东西，公司也越来越开放，组织的边界在不断被打破。组织形式就是一个公司的DNA，办公模式就是一个公司组织形式的外在。组织形式的一个体现方式就是人与人交流与连接的方式，强调灵活、开放和协作，打通企业内部与外部的交流。人与人之间的连接最主要的影响因素是见面及交流的次数，大片的开放办公场所，每个人都能看见彼此，也增加了人与人之间的连接；同时非正式即兴的交流空间，最大限度地促进了沟通并增加了信任。

在GE北京科技园设计中彻底取消格子间的大胆尝试，与其说是设计创新，不如说是新型的公司的组织架构带来的新的办公模式，新型的办公模式带来的

图 1-46 办公空间内的静音电话亭

图 1-47 员工服务自助终端

办公空间的变革。近年来，创新型的跨国企业无疑都在进行打破格子间的实践，这代表着公司内平等的组织关系与架构。

（四）共享、弹性、办公空间具有快速改变的弹性

“共享经济”就是打破现有物品所有权和使用权分离的概念，创造更多的共享、分享、交换和交流，互联网使这样的分享逐渐变为可能。总部内部共享办公空间就是针对弹性时间工作的员工设计的，当他们需要共同协作的时候，他们会选择走到为团队提供合适服务的地点，一起工作。可以根据工作任务和工作行为，对办公空间进行自由组合和分配，临时构建成不同的空间形式，承载不同的工作场景，这意味着你永远不知道谁会坐在你旁边，这个恒定的变化让同事们相互连接，而且互相启发。

谷歌总部“山景城”为了适应现代社会的快速变化，被打造成一座“随插即用型建筑”，依靠一些可轻松移动的轻质区块式建筑结构，一个团队的办公室可以装载至车辆后转移至总部的另一位置，整个空间变得像一套巨大的家具组件，可以自由组合。

GE北京科技园则提供了3000个共享的工位，打破了部门界限，同时避免弹性办公员工对办公空间的浪费，是目前国内最大规模的集团内共享办公空间。创新型的大型跨国企业，针对市场的变化迅速反应是企业发展的重要保证。部门的扩张、合并、裁撤的速度大大加快，共享办公空间的建立、部门界限的消除，使企业办公空间快速变化以适应发展。

（五）园区化、社区化、微型城市化与综合社区服务的建立

1.新型总部园区化、社区化、微型城市化发展特点

总部对于企业的意义已经不仅仅是一个办公的地方，更多是一个体现公司文化特色和提高员工工作激情和幸福感的载体。近年来，跨国企业总部建设一个明显的特征是园区化，从命名就可以看出发展的特点。GE北京科技园名为GE Beijing Technology Park，苹果的新总部则为Apple Campus，谷歌的“山景城”叫作Mountain View，除腾讯新建设的新总部为高层建筑外，大部分新型总部均采用园区化的总部模式。这和跨国企业很多研发工作需要在园区内进行大量的新产品的实验研发有关，比如正在进行研发实验的无人驾驶汽车等。

跨国企业总部除了满足基本的办公需要外，也是汇聚各部门的服务和生活配套设施中心，从内部因素看，配备多元化的高端办公服务是企业必须提供的配套功能，GE北京科技园就提供了除工作之外的就餐、休息、培训、会议、交流、娱乐、健身、医疗、母婴室、咖啡厅、超市、展示等诸多空间。蚂蚁金服总部除了一天提供四次餐食的食堂，还拥有18家不同业态的餐饮门店、24 小时营业的便利店和随处可见的自动咖啡机，另外配备健身房、邮局、超市、花店等，浓缩了一个社区乃至一座小城镇的功能。

总的看来，总部建设的园区化特征明显，园区内功能社区化、园区模拟生活社区与复制部分城市功能是近年企业总部建设的发展趋势。

2.国际化的综合实施管理服务

除了传统意义上的房屋维修、养护、清洁、保安外，在这些大型跨国企业中，物业管理的内容已拓展到物业功能布局和划分、市场行情调研和预测、物业租售推广代理、目标客户群认定、工程咨询和监理、通信及旅行安排、智能系统化服务、专门性社会保障服务等全方位服务。

（六）智能、互联——全球或区域中枢的建立

跨国公司总部与分布在世界各国的分公司、子公司之间需要建立起一个迅速有效的信息传输机制，以解决员工之间信息交流、协同作业的问题。远程视频会议系统无疑可以解决跨国公司的信息传递问题，通过清晰的视频画面和流畅的音频连接使跨国公司员工之间可以达成无障碍对接，帮助企业及时把握市场动向，提高全球不同区域团队协同工作的效率。

物联网时代服务管理和用户需求都能实现远程操

作与快速对接，让智能系统自助处理各种日常事务。手机是园区工作生活的“遥控器”，门卡、会议室预定、网络、打印、付款、娱乐、健身、医疗均集成于手机上。智能手机控制模式和系统集成最大化的发挥资源整合管理和人机交互功能，包括监测空气质量、智能化设置空调系统运行方式、自动关闭空调系统节省能源等。

随着科技和互联网的快速发展，高智能化的全球中枢的建立，园区内部智能化的运营和园区办公物联网的建设是近年办公园区发展的重点之一。

（七）绿色到健康

2014年国际建筑师联合会前任会长路易斯·考克斯（Louise Cox）在论述“中国当代建筑设计发展战略”时强调，可持续建筑和环境问题是设计的根本。在今年总部园区的规划设计中，我们看出一个重要的趋势是，办公空间从关注环境进一步延伸到关注职员的身心健康，美国LEED绿色建筑认证或者国内的绿色建筑认证成为标准配置，同时为员工提供良好的工作环境、良好的健身场所、健康的饮食等是目前大型企业追求的新目标。

美国Delos（得洛斯）实验室近年来则立足于医学研究机构提出了Well房屋标准，其中包含了和人身心健康相关的七大要素，在这七大要素中也包含了105项标准。Well是一个基于性能的系统，它测量、认证和监测空气、水、营养、光线、健康、舒适和理念等建筑环境特征。

GE北京科技园提供了窗明几净的健身场所，种类多样、健康的餐厅，办公建筑均为薄板，有良好的通风日照，同时在园区内提供了大型的核心绿化。我们看到，苹果、Facebook总部均是公园与办公空间融合，把办公室打造在公园中，阿里巴巴总部办公楼基地位于杭州郊外西溪湿地国家公园，目标是要与自然环境共生，并保留了作为地块特征的水源及绿地。

未来办公提供的是一种健康的工作生活方式，一种宜人的工作场所，通过引入Well Building健康办公理念，通过各种技术手段，切实保证员工们的健康，振奋他们的精神，提高人们的效率。

四、结语——不仅仅是标志性

“适用、经济、绿色、美观”是国家确定的新建筑方针，它契合了如今能源世界对办公环境呈低碳化、数字化且健康智能的新要求，旨在改善并提供办公环境的安全、健康、便捷、高效、舒适，重在提高办公人员的健康工作指数。面对企业结构多元化以及办公行为多样化的趋势，以公共交流空间为核心，就需要建立一个个具有较强的向心凝聚力和向四周辐射的办公空间。不仅以共享空间、流动空间形式作为总部精神场所，同时作为建筑的物质、视觉和社会中心而能唤起建筑师的创作意识，自觉地提升多样化办公的企业凝聚力。具体讲，从设计上要：体现人工地景与天然地景的融合，实现品质高端且充满活力的企业文化，展示绿色科技和智能相结合的生态景观空间等。

一般认为，新建总部大楼是展示企业核心竞争力和企业形象的一种方式，更关注建筑的标志性，但有时并不尽然。2013年苹果和亚马逊新总部的设计方案，就接连引发了对比与质疑：“这些建筑已经不单纯是公司总部大楼这么简单，它们是纪念馆，映射了如今科技不可阻挡的统治地位以及硅谷在世界经济和文化领域的霸权。”德国《明镜》周刊的一篇文章这样评论。

随着中国经济的快速发展，中国已经成为国外企业投资的一片热土，跨国公司大规模地涌入中国开拓市场，这是中国产生总部办公的一个外部因素；另一方面，中国本土企业也在快速成长，企业的规模和业务范围迅速扩大，这为总部办公准备了丰厚的企业资源。

信息时代的快速发展，从很大程度上改变了办公模式：不再是传统的一个人埋头苦干，未来注重的是团队间的相互合作，信息的快速交流。因此，多种智能、健康的活动、休息空间的引入，是信息技术的发展及企业管理模式的转变等因素推动的结果。

办公文化的变革，办公需求的变革是办公场所变革的原动力，在全新的时代，我们推测办公空间未来将是：平等与交流、共享与变化、智能与健康、社区与服务的有活力、强化交流、引发创新的办公社区生态圈。

注释

1）“独角兽”企业原本是美国风投界的术语，用来描述初创10年内估值超过10亿美元的公司。据CBinsight统计，2013年至2017年7月底全球估值超过10亿美元的独角兽企业共有214家。其中美国仍遥遥领先，有127家，中国排名第二，有59家。在独角兽企业名单上排行在前的企业有Uber（交通）， Airbnb（住宿）， Palantir（大数据）和Snapchat（社交媒体）等。

2）美国FM（AssociatedFactoryMutualFireInsuranceCompanies，或简称Factory Mutuals）Global创建于1835年，经过180多年发展，其业务范围已涵盖保险、认证、消防、标准的研发和设计等诸多领域，目前在世界保险公司的排名为24位。FM Global总部位于美国罗得岛州，具有独特的针对财产损失和商业连续性的保险产品和风险管理手段，并在180多年来一直专注于工业资产的安全，与其他保险公司不同之处首先在于，承保人的核保主要是基于风险工程师系统专业的风险工程评估所提供的精确的风险数据和信息，而不是以精算评估为基础。该公司的另一大特点在于它的互助制，它的保险客户就是它的股东，被保险人既是其客户也是公司的所有人，年底部分盈余保费以分红形式返还给被保险人，这种做法一直沿用至今，其形式为现在的FM Global成员奖励机制。共同的利益使他们能够在风险改进、减少损失方面进行合作，并通过相应的机制形成一种互动式的良性循环。FM Global专业技术来自于他们的世界级的研究中心和其中的科研人员实施的研究和测试，这种独一无二的创新型机构，为他们的工程师提供了不断更新的创新标准和规范，来支持和保证他们持续地为客户提供最好的服务。FM Global集团同时拥有一家认证机构FMRC，它是一个独立于保险企业之外的实体，它唯一的使命就是为与防损相关的材料和设备提供“世界一流”的认证，这种认证支持工程队伍为客户提供更加可靠的防损策略。

项目信息及版权所有

GE北京科技园

1）GE北京科技园使用方、投资方、设计团队及顾问团队等：

最终用户：通用电气医疗（中国）有限公司

最终用户负责人：

段小缨：通用电气公司中国区CEO

Jeffery Sommer：通用电气医疗集团亚太区供应链总经理

张配：通用电气医疗集团亚太区房地产部总监

王天翼：通用电气医疗集团设施部项目经理

建设单位：坤鼎投资管理集团股份有限公司（证券代码：833913）

设计总承包单位：北京建院约翰马丁国际建筑设计有限公司

建筑主体方案、初步设计、施工图设计：北京建院约翰马丁国际建筑设计有限公司

实验楼方案、初步设计、施工图及室内设计：北京建院约翰马丁国际建筑设计有限公司

办公研发部分室内概念设计：AECOM

办公研发部分室内设计施工图：北京建院约翰马丁国际建筑设计有限公司&北京建院装饰工程有限公司

实验楼部分工艺设计：北京建院约翰马丁国际建筑设计有限公司

实验楼部分室内设计：北京建院约翰马丁国际建筑设计有限公司

园林与外线方案、初步设计、施工图设计：北京建院约翰马丁国际建筑设计有限公司

弱电方案、初步设计、施工图设计：北京建院约翰马丁国际建筑设计有限公司

设计总指导：张宇（北京市建筑设计研究院有限公司副董事长、全国勘察设计大师）

项目总负责人：朱颖

设计团队：1.邹雪红 2.朱颖 3.葛亚萍 4.鲁晟 5.朱琳 6.Cory Ticktin（AECOM）7.周彰青 8.张涛 9.李荣辉 10.杨超 11.田玉香 12.张胜 13.章伟 14.常青 15.张建 16.赵伟 17.彭晓佳 18.李轩 19.刘昕 20.陈瑜 21.杨一萍 22.杨春红 23.刘佳 24.朱义秀 25.傅九思 26.付晓琳 27.张爱强 28.张广宇 29.袁茵 30.陈威 31.刘禹希 32.朱伟新 33.白楠 等

顾问团队：世邦魏理仕CBRE

项目负责人：张利民 梁煜

施工单位：中建二局第四建筑工程有限公司

监理单位：北京海建工程建设监理有限责任公司

2）设计时间：2012—2016年

3）项目取得了LEED认证并通过了FM Global工程风险评估

4）获奖情况：

2017年BIAD优秀建筑工程设计（公共建筑）一等奖

2017年北京市优秀建筑工程设计项目二等奖

2017年全国优秀工程勘察设计行业奖优秀建筑工程设计项目一等奖

5）图片版权：除部分图片由通用电气（GE）公司、坤鼎集团、AECOM提供外，其余图片版权均属于北京建院约翰马丁国际建筑设计有限公司

6）竞赛时项目名称为“通用电气医疗中国研发试产运营科技园”，实施过程中GE中国决定将之前分散于北京各个办公地点的包括航空、发电、医疗、油气、可再生能源、运输、能源互联、研发及数字等不同的业务部门与GE医疗集团全球的研发设计、中国区的运营管理及部分销售统一整合在一起，打造其GE中国北方区的运营和研发总部，并更名为“GE北京科技园”（GE Beijing Technology Park，简称BTP）

7）撰文：朱颖 周彰青 邹雪红 常青 赵伟 彭晓佳 杨一萍

贰记 Note Two

现代主义实践

Modernistic Design Practices

某科研中心、北京国际旅行卫生保健中心、琥珀天地、
月季洲际大会会议中心设计实践

A Research Center, Beijing International Travel Health Care Center, Hupo Tiandi
Neighborhood, and Conference Center for WFRS Regional Convention

建筑要随时代而发展，中国的现代建筑应同工业化社会相适应。
建筑设计要建立在合理分析问题和解决问题的基础上，
“形式追随功能是不变的真理”。
积极采用新材料、新结构，发挥新材料、新结构的特性，
自由平面、自由立面、屋顶花园、底层架空、横向长窗，
“新建筑五点”是新材料、新结构带来的建筑特征。
我们努力实现表现手法和建造手段的统一，我们关注建筑形象的逻辑性，
我们采用灵活均衡的非对称构图、材料与体型有机结合的设计手段，
我们关注易建性，但这一切最终目标仍然是创造有特质的建筑形象。

As architecture develops with the times, modern Chinese architecture should be compatible with an industrialized society. Architectural design should be based on reasonable analysis and solution of problems, “Form follows function” is an immutable truth. We actively adopt new materials, new structures and bring their characteristics into full play. Free plan, freestanding facades, roof garden, stilt building and horizontal strip windows represent the "Five Key Elements to New Architecture”, which are the architectural features brought by new materials and new structures. We strive to achieve harmony of expression skills and construction means. We focus on the logic of architectural image. We introduce flexible and balanced asymmetric compositions and the design approaches combining materials and forms in an organic manner. We are concerned about buildability, but the ultimate goal is always to create a distinctive architectural image.

建筑与环境对话：某科研中心

（一）方案的生成

某科研中心是一个小型项目，主要功能是高科技研发用房及相关配套办公。

由于项目附近都是山体，可建设用地范围较小，因此建筑的总平面布局和体量成为设计团队关心的重点。建筑用地内部的东侧是高起来的山体，以岩石为主，用地内部西侧为一条自南向北沿山势而上的内部道路，紧临道路西面为一条天然形成的泄洪沟，业主在这里挖了一片池塘，并做了些水上的廊桥。业主要求的建筑规模在城市里的确是个小型建筑，但在基地附近，由于是半山腰，周边同等高度没有建筑，因此这个建筑绝对是个庞然大物。设计团队在总图设计阶段做了多方案的比较，也同业主做了多轮的探讨，决定把建筑物进行体量拆解，形成一大一小两个建筑，两个建筑用连廊相连，这样拆解后降低了建筑的体量，使其尽量与周边协调，同时拆解后的体量适于总图排布。

最后就形成了中间主楼和其东南角配楼相搭配的方式，主楼为4层，配楼为3层。主楼和配楼共同围合形成入口广场，广场与西侧的池塘互相映衬，形成了较强的场所感。由于主楼为东西向较宽的建筑，其西侧虽避让了上山的道路，但在空间感受上仍然不是很通畅，设计团队于是把建筑首层西侧的部分处理成架空的灰空间，这样首层的灰空间即成为室内门厅的室外延伸，同时也和西侧的池塘进行了很好的对话。更为重要的是灰空间也成为西侧上山道路的扩展，使得整个培训中心自南至北的视觉轴线和通廊得以保留和强化。而某科研中心恰巧是这一轴线上的最重要的节点。这一设计既凸显了某科研中心的重要性，也对培训中心的院内空间格局进行了清晰的划分和梳理。

（二）建筑形式的表达

建筑的立面设计是设计团队所关心的另一个重要问题。配楼3层高，与周边的山体高度一致，不会凸显。但主楼高达4层，最上面一层从视觉角度来看略显突兀，与周边的山体绿化不是很协调。如何利用立面设计弥补这一缺点，是设计中的重点。同时，某科研中心是科研办公性质，什么样的立面形式既能凸显建筑性质，又能体现项目的独特性，也成为研究的课题。

图 2-1 西南侧效果图

图 2-2 西侧实景

图 2-3 北侧实景

图 2-4 东侧实景

图 2-5 入口细部

图 2-6 西侧转角

图 2-7 入口广场夜景

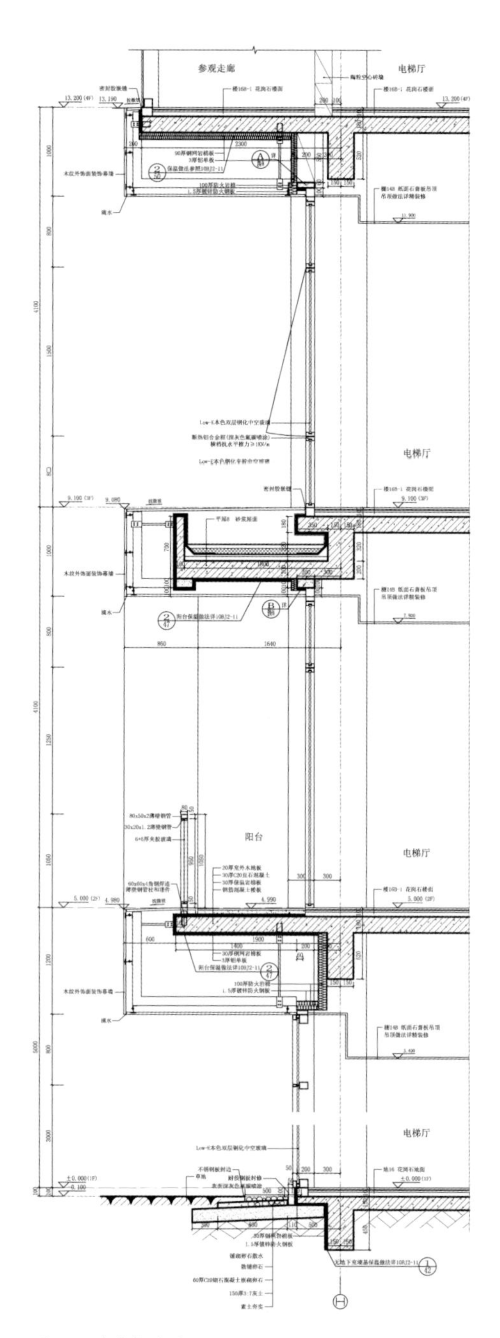

图 2-8 外墙构造详图

图 2-9 西南侧夜景

最后设计手法还是回归到了体量的拆解上。从西侧的首层架空延展到整个建筑，设计团队将建筑物的首层都处理成为玻璃幕墙的立面，这样既能保证首层的采光，也能利用玻璃幕墙中周边植被的影像与环境融为一体。建筑2层和3层则采用变换材质的方式形成了水平体量处理，这一处理在主楼西侧和配楼南侧的端头形成了与DNA双螺旋结构相近的S形，整个建筑体量的尺度降了下来。建筑的 4 层，设计团队依然采用了玻璃幕墙的手法，希望能透过玻璃幕墙的特性与周边环境取得一致，降低顶层的突兀感，这一处理方式最终被证明是十分有效的。

对建筑立面的拆解和构成方式的研究仅是在建筑立面的物理层面所开展的探讨，但建筑立面的表意性即某科研中心的形象特质还没有落位。设计团队提出把支撑架空部分的结构调整成DNA字母的形式，这一想法对整个建筑立面起到了画龙点睛的作用。DNA字母做到了完美的形式和结构的统一，同时通过这种直接的表意处理给建筑增加了当代艺术气息，并直接点出了项目的内容特质。这种文字与立面的结合处理指引了一条建筑立面处理的新途径，最后的效果也是十分突出和震撼。建筑立面的实体部分材料采用仿木的千思板，材料的颜色和质感与实木相当接近，与环境也能更好地融合。

（三）结语

最终的建筑效果完美地体现了科研办公建筑清晰规整的特性，4层的玻璃幕墙映衬着蓝天，仿佛和天空融为一体。DNA三个白色字母有力地承托着木色的水平向建筑主体，整个建筑轻盈、明亮、干净。这个崭新的建筑给这片区域带来了清新的活力。

图 2-10 西北向效果图

城市细胞：北京国际旅行卫生保健中心

（一）项目背景与概况

项目用地为北京国际旅行卫生保健中心自有用地，用地内现存一座建于20世纪70年代的4层业务楼及一些附属平房，目前均在使用中。由于业务发展，现有业务楼已经远远不能满足使用要求，因此需要在拆除旧楼的基础上建设新的综合业务楼。

项目用地北侧临城市道路和平里北街，东侧和南侧紧临和平里医院，西侧与住宅小区由一条规划路相隔。用地北侧临街面正对一过街天桥；东侧为和平里医院5层门诊楼，距离本项目用地约2米；南侧为和平里医院临建平房，西侧隔规划路是一栋12层住宅楼，距离本项目用地约10米。

项目用地现状情况复杂，且限制性条件较多，为设计的进行和实施带来了挑战。

（二）方案的生成

鉴于项目用地现状及用地规模条件所限，设计团队在总平面布局中将建筑物紧凑布置、整体设计，尽量留出更多的室外空间，以满足室外景观环境与消防的需要。

最终实施的方案建筑位于基地正中，建筑物形态规整，呈长方形，紧凑集中，建筑主朝向为北向，长边正对和平里北街。用地西北角设置主要出入口，西南角设置消防紧急出入口。地下车库出入口位于主入口近旁，日常使用车辆直接进入车库。消防车道环绕建筑北、东、南三边，扑救面涵盖建筑的两个长边。

（三）建筑形式的表达

由于建筑的体量十分规整，且规模不大，设计团队在建筑形象设计中确立了设计原则：建筑造型设计上力求突出建筑的整体性，讲究造型比例适度、空间结构明确美观，强调外观的明快、简洁。

由于建筑形体严整、方正，整体造型以方形为主。为突出建筑整体性，外立面设置规整的条形窗，条形窗在建筑表面形成均匀的竖向立面肌理，使建筑立面完整统一、优雅简洁、端庄大气。

为了使方正的建筑立面不显单调，北侧沿街主立面通过局部微妙的退台处理，在建筑表面形成舒缓的

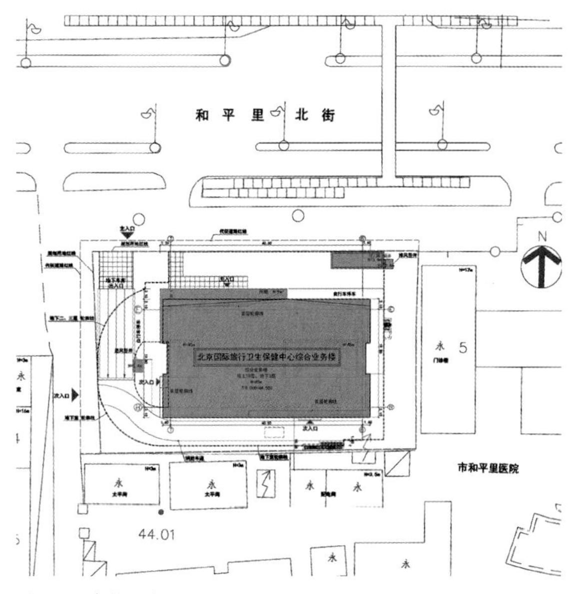

图 2-11 总平面图

图 2-12 北侧实景

图 2-13 东南侧实景

图 2-14 公共服务大厅效果图

图 2-15 入口中庭效果图

凹凸起伏，与竖向线条相配合，使建筑产生向上生长蔓延的独特视觉效果，仿佛赋予其生命。两个斜向面使得建筑产生迎向东西两侧的效果，极大地增强了建筑的标志性和可识别性。

建筑北侧1~2层利用局部玻璃幕墙和挑出的雨棚，形成通透大气、两层通高的建筑主入口。突出了建筑的服务性和开放性。同时，入口幕墙造型配合雨棚形成对主入口广场的环抱和围合，充分体现了建筑的亲和力，并形成了极具人性尺度的入口空间。建筑入口内部的二层挑空空间和联系上下层的自动扶梯，通过通透的玻璃实现了与外部自然环境的沟通和融合，可极大地改善体检和来访人员的心理感受。

建筑顶部为屋顶花园，采用由建筑立面向上延续的构架和百叶进行开放式的围护，形成了优雅、舒适、宜人的室外空间，同时配以小型庭院景观和绿植，既符合国家节能环保屋顶绿化的倡导，又给建筑自身提供了更丰富的多功能空间。

（四）建筑内部功能及流线组织

本项目主要使用功能有各类体检、实验、办公等，其日常业务使用具有体检种类多、瞬间人流量大、人员流线复杂且具有不确定性的特点。而由于用地条件的限制，本项目单层建筑面积有限。因此，如何将各类体检功能有序排布，合理组织竖向和横向交通，迅速疏导人流，避免人员瞬时拥堵，是本项目地上使用功能布置的设计要点。

因此，在平面布局中，将建筑的主入口及共享大厅与接待收费功能相结合，布置于建筑北侧。与共享大厅紧密结合组织竖向交通，通过自动扶梯、开敞楼梯、位于两个核心筒的楼梯、电梯等竖向交通形式，将体检人员迅速运送至各体检楼层。

核心筒位于建筑居中位置，围绕核心筒及建筑中部的共享空间布置体检房间，视线畅通、流线简单明确，有利于快速、有序使用。

本项目地下部分在用地范围、层高、层数等方面受限，地下室的功能既有车库、建筑运行所必须的机电设备用房等，还有后勤管理、餐厅、厨房、人防等，功能齐全。在设计中，将地下各功能紧凑、合理、有效地排布。

（五）结语

现代城市中一些建筑通过新、奇、怪的处理方式来博人眼球，宣扬自我，但却往往忽视周围环境，使周边城市空间支离破碎。北京国际旅行卫生保健中心通过紧凑合理的方正体量，比例严谨的立面肌理，把周边城市空间整合起来，并在整合后的城市空间中，恰当地描述自身的存在感，这种强调建筑是城市组成细胞的现代主义设计手法是需要一直坚持的。

舞动的神龙：琥珀天地商业设计

（一）项目背景与概况

北京远郊区县各卫星城起着疏解中心城区交通压力和分散城市功能的作用，对于北京城市长远规划和合理发展具有十分重要的意义。昌平区作为北京北部重要卫星城区，其地理位置和周边环境都彰显着天时地利人和，现在乃至未来都是承担居住、就业、高新产业、文化产业等重要城市职能的地区。

回龙观居住文化区是昌平区较早开发的一个居住区。南距德胜门约16千米，北距昌平主城区约18千米，东边是天通苑居住区，西侧紧临京藏高速公路，距沙河镇不远，现已形成较为完善的居住和配套设施。地铁13号线贯穿其中，使得居住在此的人们可以方便地进出中心城区。

本商业项目具有较好的地理位置，位于回龙观中轴线和区域主干道——十里长街（即回龙观西大街）上，是该居住区通往京藏高速公路和地铁13号线的必经之路。地块位于十里长街西部，南临十里长街，北接风雅园南街，西至周庄西路（食街）与华联商厦相望，东临邮电局和云柏鞋城，是回龙观区域商业最为集中和商业氛围最为浓厚的地区。

（二）方案的生成

本项目地块东西长将近200米，南北宽70米，呈狭长形态。200米长的面宽面对城市街道，使得这块地成为商业开发的福地，也给设计带来不小的挑战。

我们要解决的一个关键问题是如何让160米长的建筑不显得冗长，从而减少对城市的压迫感，并营造良好的商业氛围和城市外部空间，让人们愿意进入。与此同时，还要通过建筑语言表达当地的地域文化，成为一个能够自然融入该区域的合适的建筑。

最开始，我们尝试一字形的板楼形态，在不影响北侧住宅小区日照的条件下，南侧留出较为开阔的商业广场。但随之而来的是一字形的建筑横贯地块东

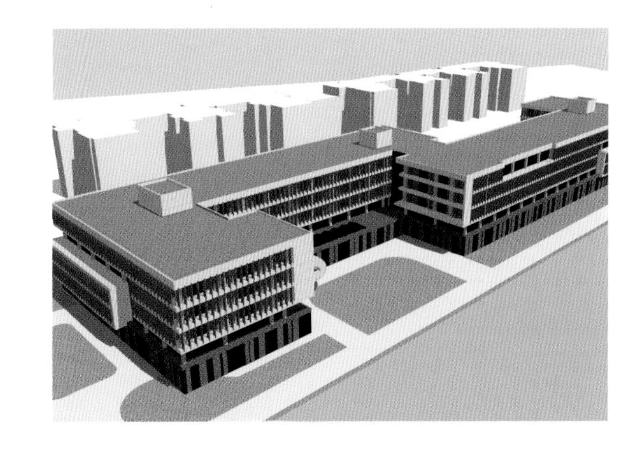

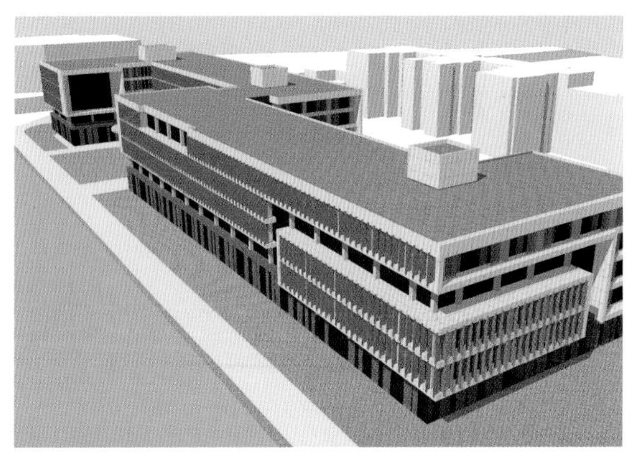

图 2-16 概念生成示意（组图）

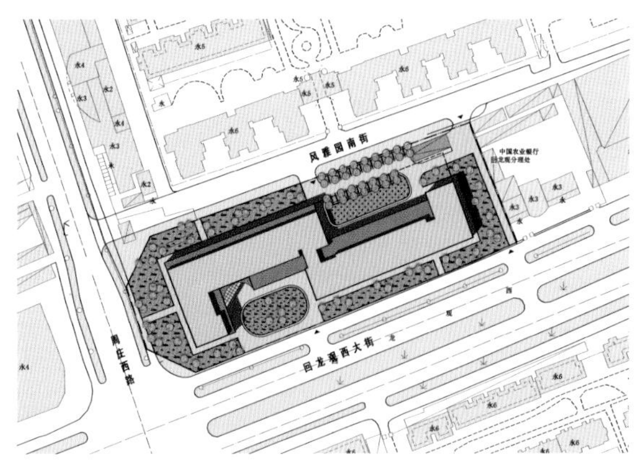

图 2-17 总平面图

图 2-18 东南侧实景

西，尺度过长，并没有亲近感。

之后，在做了大量的体块分析以及对当地历史文化的研读与探究后，我们得出了一个较为理想的布局形态，即两个“L”形体量相互扣接，形成一个连续的“S”形造型。

回龙观这个地名源自当地的一个村庄回龙观村。该名可以追溯到明代，源自明代皇帝十三陵谒陵，相传此地就有一处道观，名为玄福观，是皇帝谒陵回京途中小歇之处，久而久之，此观便被叫作回龙观。此地可以说有着深厚的帝文化和龙文化。建筑的灵感便来自于中国的龙。针对地块狭长的特征和对场地的认知，我们运用“S”形的建筑造型隐喻出龙生动飞舞的动势，能够很好地融入当地的历史文脉，使其成为回龙观地区又一具有特色的建筑。

在这种特征的地块上，“S”形体量自然围合出南北两个半开放的庭院空间。南侧庭院向城市主要界面开敞，具有很强的城市性特征，形成引导和激发商业互动行为以及人们交往的主要广场空间。北侧庭院相对静谧，与后面的住宅小区毗邻，使得建筑与住宅楼之间形成一个缓冲空间，最大限度降低了新建建筑对住宅的影响。其主要功能是组织用地内停车和人们休闲小憩以及建筑2至5层办公人流的流线。

“S”形造型形成了建筑体量的转折，从而巧妙避免了160米尺度过长带来的冗长与压迫感。建筑不再呆板，显得更加活泼灵动，符合商业办公建筑的性质，并且在形式上契合了当地的地域文化。

（三）建筑形式的表达

在建筑形式的处理上，我们为了做到建筑的节能，尽可能地不使用大面积玻璃幕墙，而通过其他手法体现建筑与外部环境的交融和其轻盈通透之感。

建筑高度18米，共5层，水平延展于地块之上，平和而自然。在竖向呈现三段式：1层的基座层，对应首层的商业功能；2层玻璃材质的虚化处理形成退后的层次；之上架起建筑的主体，对应2至5层的办公功能。这样的处理使得建筑更加轻盈漂浮。

图 2-19 南侧沿街立面实景

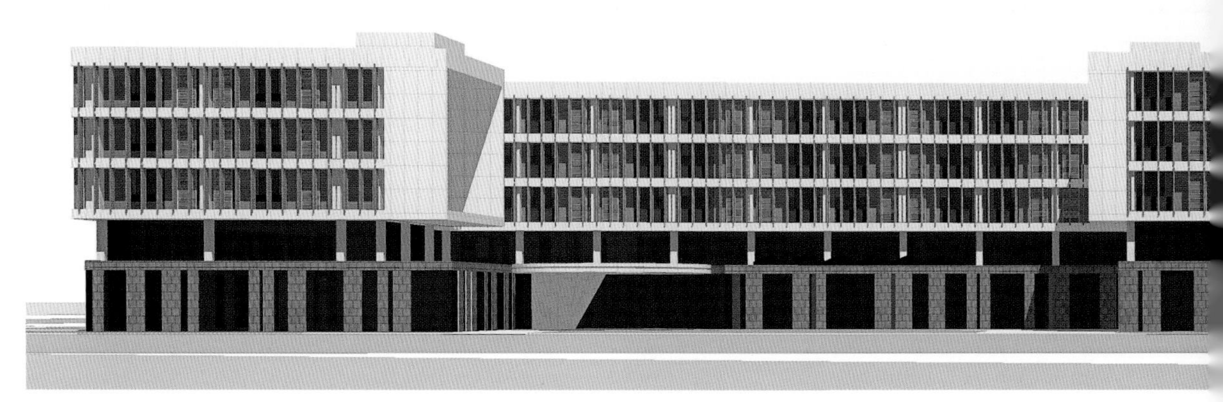

图 2-20 南侧沿街立面效果图

琥珀天地AMBER
阿里斯顿

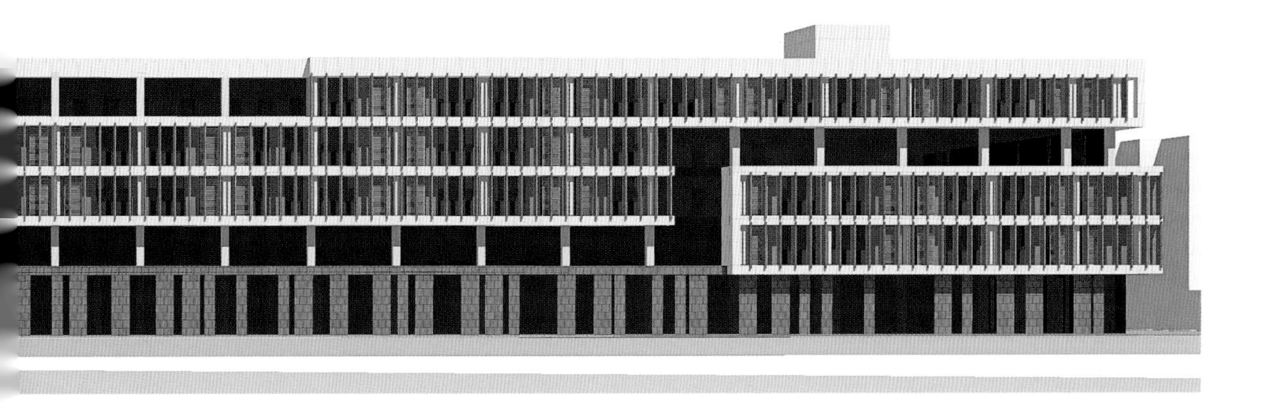

图 2-21 南侧立面转角

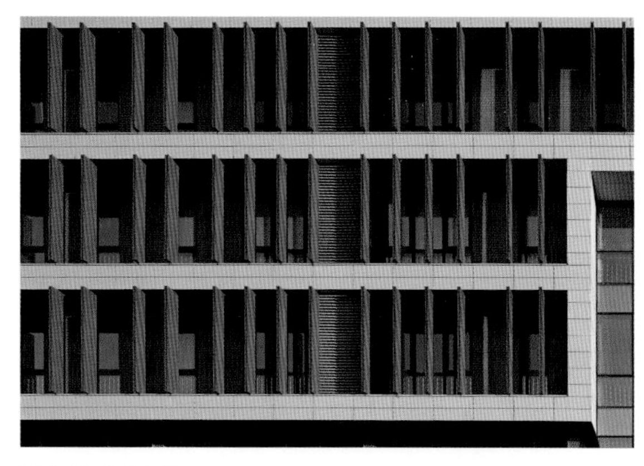

图 2-22 立面细部

图 2-23 入口广场实景

图 2-24 东南侧效果图

图 2-25 夜景效果图

坚实的基座之上通过体块穿插和错动手法在原有秩序下增加一定的变化感和复杂度，并通过双层表皮的运用使建筑真正的外维护体系退隐于水平梁板和竖向百叶构成的表皮之后，形成建筑室内空间与外部环境的柔和过渡，视觉上显得更加轻盈通透。体量端部面向庭院的部分被处理成玻璃幕墙的形式，可以与外部庭院空间构成良好的内外互动关系，并且丰富建筑的形式表达。

首层基座运用深灰色花岗岩石材墙面，承托于基座之上的主体建筑围护结构为灰色面砖和有韵律的开窗，在其之外的第二层表皮则由米黄色洞石石材和连续竖向木质百叶构成。基座和主体在色彩上形成对比：深色石材强化基座的稳重感，灰色面砖的运用直接源自北京传统民居上的灰色砖墙，而米色和黄色则代表了皇家文化，作为外表皮展现，与此地历史文化一脉相承，同时表现主体的轻盈感。

由于建筑采用分体式空调，如何把空调室外机隐藏起来从而避免对建筑外立面的破坏，就成为建筑外层表皮生成的重要问题。竖向木百叶的处理较好地解决了这个问题，同时形成较为现代的建筑形式，并且起到了很好的遮阳作用。

（四）建筑功能与流线组织

设计结合周边道路的具体情况，本着“分流、高效、灵活”的原则，对交通流线组织根据不同功能区和不同交通方式做出合理设计。

行人主要出入口设置在基地南侧商业人行主广场处，直接与城市人行道相临。在基地北侧设置非机动车出入口（兼作人行出入口），直通地下自行车车库。

机动车出入口安排在基地北侧风雅园南街上，与大量商业人行流线尽可能分离。基地的东侧设有南北向车行道，贯通本地块的车行流线，并兼作消防通道。

人流分为商业人流和办公人流。商业人流主要通过建筑南侧人行路及商业广场进入建筑内部。办公人流主要通过北侧绿化人行广场进入建筑内部。

建筑内部功能和流线的组织本着功能互不干扰和便于运营管理的原则进行竖向分区：1层为对外营业的商业店铺；2至5层为办公。1层有独立门厅和通往楼上各层的竖向交通，与1层商业在功能和流线上互不干扰。

建筑主要采用5.7米×7.9米、5.7米×8.6米柱网系统。5.7米的柱距可以停放两辆机动车，同时适合底商的分割布局，还可以满足2至5层办公空间的不同开间办公模式，可分可合，增加了建筑单元出售和使用的灵活性。

地下1层平面布置机动车车库、机电设备用房、自行车车库、戊类库房等用房。机动车车库可以满足99辆车的停车需求。机电设备用房包括锅炉间、热交换机房、变配电间、弱电机房、中水机房、消防泵房、消防水池间、送风机房、排烟机房等。

首层平面布置商业店铺，有1个主要出入口和3个次要出入口，可以方便进出商业内街。每一个独立店铺又有直接对外的出入口。为了满足上面各层的办公使用需求，首层分别设置3个办公门厅，与商业流线完全分开。每个门厅都设置专门的电梯通向各办公层。另外，在建筑东北侧设置整个建筑的消防控制中心，有独立对外的出入口。

2至5层为办公功能。办公空间依靠内走道串联起各个办公室，形成小开间办公。业主也可根据实际需求进行重新拆分组合，形成大开间办公和大小开间组合式等多种办公模式。

（五）结语

在此，我们希望通过这个项目与开发商形成共赢。开发商的决策和投资为该区域创造了一个集商业和办公为一体的综合性建筑，给回龙观地区注入了新的活力和商机，并且得到了应有的投资回报。对我们来说，通过合理的设计和建筑语言的表达，展示了我们对回龙观文化居住区的宏观理解和微观表述，创造具有地域特征的现代建筑以及传递节能环保的设计理念，为昌平新城的建设与发展添砖加瓦。

图 2-26 西立面效果图

形式自由空间：月季洲际大会会议中心

（一）项目背景与概况

世界月季洲际大会是由世界月季联合会主办，各成员国承办的全球月季界的最高级别盛会，每三年举办一次。世界月季联合会由世界各国月季协会组成，现包括英国、美国、法国、德国、中国等在内的41个成员国。被誉为“月季里的奥运会”的2016年世界月季大会为唯一一次四会合一的盛会，即世界月季洲际大会、世界古老月季大会、中国月季展和北京月季文化节同期在北京大兴区举办。承担大会主要活动的会议中心位于大兴区魏善庄镇“北京城建・北京密码”的4−2#楼。

北京密码项目位于北京南中轴线六环外西侧，大兴区魏善庄镇，东距京开高速路7千米，用地东至龙发大街，西至龙旺大街，南至龙江路，北至后查路。地块东侧为国家新媒体产业基地拟建用地，此地块东侧为南中轴路，南侧为国开会议中心和国家开放大学魏善庄校区等。

项目总建设用地规模14.27万平方米，总建筑面

42.54万平方米，由6栋7层办公楼、1栋多层独立式商业楼4-2#楼及地下车库组成。其中项目配建的4-2#楼独立商业楼作为月季大会会议中心，是2016年世界月季洲际大会的主会场。

（二）设计原则

会展中心的总平面图和建筑轮廓在之前北京密码项目整体规划设计中已经被确定下来。得知被作为2016年世界月季洲际大会主会场后，该会展中心的设计集中到了两个方面：立面设计和内部功能设计。设计团队在深入了解使用需求后确定如下设计原则：建筑设计要遵循开放式、可持续发展的设计理念，提供高适变性建筑空间。

（三）功能的灵活可变

在功能设计方面，设计团队力求合理定位垂直交通（楼梯、电梯、电动扶梯）位置，为使用者提供最大的自由空间；在建筑首层配合规划，中部留17米宽过街通廊，自然形成南北两部分商业区，通廊成为自由交流空间。利用经济实用的8.4米柱网框架结

图 2-27 东侧鸟瞰效果图

构，提供最大不定义使用功能的空间，二层局部采用网架空间结构，提供高大无柱空间，既为举行大型会议、文体活动等提供必要条件，又为未来市场变化带来的不同使用需求提供便利转换条件，减少不必要的拆改造成的浪费，实现使用利益最大化和可持续发展建筑理念。建筑西侧沿城市交通路有约80米长商业展示面，东侧与环绕的办公楼首层商业形成蜿蜒曲折的商业步行街。二层西退屋顶绿化休闲平台，以宜人的尺度为来此漫步的人们带来不一样的美好享受。在这里，舒适的空间体验超越了建筑形象成为主导。

会议中心由地上2层，地下2层组成，地下1层预留就餐区，地下2层为车库。按甲方要求：首层由南北个会展区、接待区、办公区组成；2层有600人报告厅、多功能厅、会议室、VIP接待室、休闲区、室外绿化平台休息区组成。会展期间，参会人员通过首层开阔的人行过街通廊可方便到达各功能区。行人可通过自动扶梯到达二层会议区，二层接待区高大空间明亮通透。人们平步进入屋顶休息平台，尽享室外美景。大会议厅内装修简洁明快，经济实用，突出月季大会会议主题。

考虑到会后的使用，设计团队依据甲方开发商业业态要求，布置轻质墙体就能顺利转型，不需改动结构、楼电梯等硬件设施，以开放式空间设计完美实现建筑可持续发展。2层局部采用空间网架设计，提供开敞式大空间，为大型会议提供条件，也为以后项目面对变化的市场需要提供可持续发展的可能。

（四）艺术品化的立面

建筑的立面处理是设计的重点，如何体现建筑的公众性，如何体现会展主题，如何突出建筑自身的形体特点，这都成为设计团队思考的内容。最后，设计团队确定采用双表皮的设计模式。内部表皮解决通风采光等基础功能，外部表皮解决建筑形象。设计团队力图将外表皮处理成一件艺术品。建筑主体外部环绕一个圆形轮廓，有韵律变化的折板式穿孔铝板在建筑外围形成一层帷幕，这层帷幕既有简洁、清晰的现代立体轮廓，又隐含着对植物和花瓣的喻意。通透的外表皮仿佛给建筑穿上了一层充满褶皱的薄纱，使建筑整体显得轻盈优雅。

曲线的造型配合折板式的处理，再加上穿孔的效果，使建筑在各个角度有不同的观感，现代、简约、前卫，极具创意性、个性化。

建筑立面分成两段式，首层柱廊内配通透玻璃门窗，内外空间视线通畅，吸引人们进入，达到内部商业使用功能与外部形象的统一。2层穿孔铝折板仿佛升在空中，随风飘动。其引人瞩目的独特地标造型成就了该项目的商业活动中心地位。

双层立面体系不但很好地解决了立面的造型处理，也创意性地解决了建筑的很多问题。内层解决采光、通风、保温节能、防水防潮、防噪声问题，外层采用耐久性高的乳白色穿孔铝质折板，沿主体流线向两侧展开，形成中间自然的入口的引导。铝板具有环保、抗腐蚀、防水、抗紫外线、容易清洗、耐久性好的特点，能很好地延长立面的使用寿命。

建筑的屋顶也是设计团队考虑的内容。由于周围建筑较高，且项目位于新机场附近，设计屋顶时采用可上人绿化种植屋面，既增加休闲空间，又美化空间形态。

（五）结语

本项目巧妙地利用环保绿色的新材料、新技术，将现代绿色、生态理念及高标准建筑节能技术融入建筑设计中；并充分考虑城市场所的时空表达、文化内涵，提供具有艺术美、时代特色的地标性建筑，从理念、科技到创意引领了该新区的建筑设计发展方向。

同时，设计团队坚持开放式建筑设计理念：空间的存在是恒常的，但其内容的使用却是可以变化的。建筑师不应在一次设计中定义所有空间，要能让使用者参与其间，提供高适可变的建筑空间。这可以说是对现代主义建筑的一种延展和深化。

项目信息及版权所有

某科研中心
建筑主体方案、初步设计、施工图设计：北京建院约翰马丁国际建筑设计有限公司
设计团队：1.朱颖 2.邹雪红 3.王鹏 4.朱琳 5.葛亚萍 6.田玉香 7. 周宏宇 8.许阳 9.熊进华 10.赵伟 11.张金玉 12.姜亚慧 13. 姜建中 14.薛磊 15.张爱强 16.陈晓洁
基地面积：2700平方米
总建筑面积：5730平方米
项目状态：已建成
设计时间：2011—2012年
建成时间：2012年
获奖情况：
2014年度BIAD优秀工程设计奖公共建筑类二等奖
2015年第十八届北京市建筑工程优秀设计奖
摄影/图片版权：陈鹤/北京建院约翰马丁国际建筑设计有限公司

北京国际旅行卫生保健中心
业主：北京国际旅行卫生保健中心
建设地点：北京市东城区
建筑设计：北京建院约翰马丁国际建筑设计有限公司
室内设计：北京扶桑建筑装饰有限公司
设计指导：刘力（全国工程勘察设计大师）
设计团队：1.朱颖 2.邹雪红 3.朱琳 4.王鹏 5.葛亚萍 6.周彰青 7.张爱强 8.田玉香 9.钱文庭 10.章伟 11.常青 12.赵伟 13.彭晓佳 14.姜雅慧 15.赵欣然 16.姜建中 17.迟珊 18.杨春红 19.赵阳
基地面积：3024平方米
总建筑面积：15115平方米
项目状态:已建成
设计时间：2012—2015年
建成时间：2017年
摄影/图片版权：朱有恒/北京建院约翰马丁国际建筑设计有限公司

琥珀天地（回龙观A08地块配套商业用房项目）
业主：北京智地风雅房地产开发有限公司
建设地点：北京市昌平区
建筑设计：北京建院约翰马丁国际建筑设计有限公司
设计团队：1.朱颖 2.沈枕 3.王鹏 4.邹雪红 5.曾劲 6.朱琳 7.田玉香 8.董小海 9.王琼 10.熊进华 11.姜建中 12.彭晓佳 13.薛磊 14.许阳 15.赵静远
基地面积：17699平方米
总建筑面积：27023平方米
项目状态：已建成
设计时间：2010—2011年
建成时间：2012年
获奖情况：
2014年北京建院优秀工程二等奖
2015年第十八届北京市优秀建筑工程设计项目优秀奖
摄影/图片版权：杨超英 陈鹤/北京建院约翰马丁国际建筑设计有限公司

月季洲际大会会议中心
业主：北京城建兴华地产有限公司
建设地点：北京市大兴区
建筑设计：北京建院约翰马丁国际建筑设计有限公司
设计指导：何玉如（全国工程勘察设计大师）
设计团队：1.张彤梅 2.杨林 3.李艳 4.王鹏 5.朱颖 6.胡益莎 7.梁晨 8.王晓光 9.沙鸣娜 10.董伟 11.田立宗 12.赵波 13.刘昕 14.尹鹏 15.杨春红 16.刘佳 17. 赵阳 18.姜建中
基地面积：92123平方米
总建筑面积：6640平方米
项目状态：已建成
设计时间：2014—2015年
建成时间：2015年
摄影/图片版权：北京建院约翰马丁国际建筑设计有限公司
撰文：王鹏 朱颖 张彤梅

叁记 Note Three

城市活力客厅

Reception Room of Urban Vitality

吉安文化艺术中心创作实践与运营后评估

Ji’an Cultural and Art Center Design Practice and Post-operation Evaluation

从文化设施到文化空间是当前国内外规划设计界服务文化城市建设的重要趋向。
2015年中央城市工作会议强调，
“统筹改革、科技、文化三大动力，提高城市发展持续性”，
这无疑是将文化作为城市发展主动力之一的新提法。
本记通过文化创新的个案——城市文化空间设计，
在探索城市基质、塑造城市导引型文化实践上提出构想，
关注城市文化容器、构建、运营，从而使城市文化空间要素生命鲜活、历久弥新。

The shift from cultural facility to cultural space represents a major trend the planning and design communities at home and abroad have set while they are helping constructing cultural cities. At the 2015 National City Conference, Chinese government stressed to coordinate the three driving forces of reform, technology and culture with a view to propelling cities toward sustained development. It is definitely the first time for China to propose culture as a major engine of urban development. This paper introduces some cases about how to design innovatively cultural space in cities. After exploring the fundamentals of urban space and fostering the culture where cities play a guiding role, it comes up with some ideas to pay attention to the media, construction and operation of urban culture and thus making the spatial elements relating to urban culture more vibrant with the passage of time.

尽管世界上最古老的剧场，一般认为是约建于公元前525年的希腊的索瑞格斯剧场（Thorikos Theatre），但真正意义上的现代的“镜框式剧场”则出现于1618年的意大利的帕尔玛，被称为“法尔内塞剧场”（Teatro Farnese），它被认为是以后几百年的镜框式舞台剧院的鼻祖。与西方不同，我国的戏曲虽然源远流长，但近代之前的戏曲表演场所和现代意义的剧院并不类似，直到19世纪，现代意义的镜框式剧场才被直接移植到中国。改革开放之后，1989年建成的深圳大剧院开启了我国现代大型剧场建设的序幕，以1998年上海大剧院落成为起点，北京国家大剧院开始建设为标志，国内掀起了一股建设文化艺术中心的热潮，我国各省、市和发达的县级市都开始建设演艺中心、剧场、会堂等建筑。我国进入了现代剧场建设近20年的繁荣期。

北京建院约翰马丁国际建筑设计有限公司于2009年接受吉安市文广局委托开始进行吉安文化艺术中心的建筑设计，2012年年初剧院部分竣工并投入使用，2013年整体竣工。

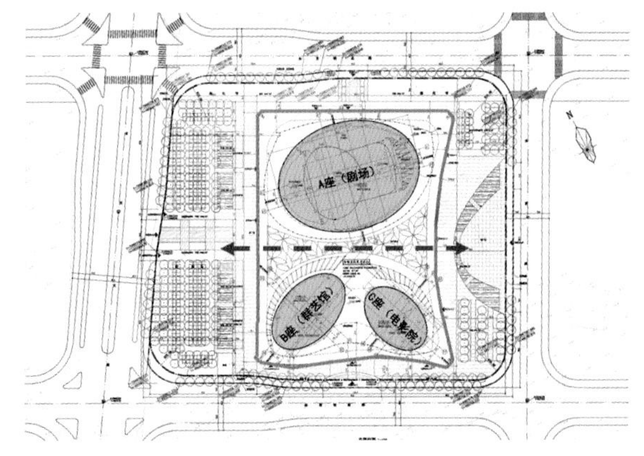

图 3-1 总平面图

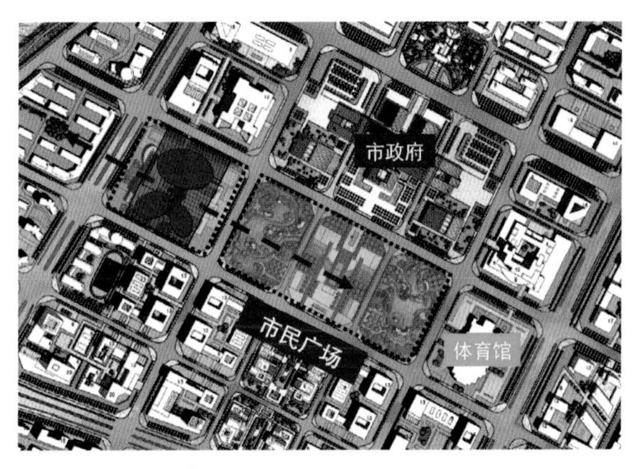

图 3-2 环境区位分析图

一、吉安文化艺术中心建设概况

吉安位于江西省中西部，是举世闻名的革命摇篮井冈山的所在地。吉安古称庐陵、吉州，元皇庆元年（1312年）取吉阳、安成首字合称为吉安。吉安自古人杰地灵，文化发达，民风淳厚，素有“江南望郡”“金庐陵”之称，更有“文章节义之邦”的美誉。从唐宋至明清，吉安科举进士近3000名，状元、榜眼、探花52位（状元17位），曾经出现过“隔河两宰相，五里三状元”的历史盛事。“唐宋八大家”之一欧阳修、民族英雄文天祥、《永乐大典》主纂解缙、宋代大文豪杨万里等一批历史文化名人先后诞生在这里，被誉为“人文故郡”。这里诞生了伟大的“井冈山精神”，被誉为“红色摇篮”。

吉安文化艺术中心选址位于吉安城南新区，项目用地西临吉州大道，北、东、南三面分别为规划中的体育馆北路、规划路、体育馆南路。用地东侧为行政中心前的市民广场。项目建设用地形状近似梯形，面积约为4.54公顷。文化艺术中心与其东侧的体育馆、东北侧的行政中心呈品字形排布，共同围合成行政中心前的市民广场。文化艺术中心总建筑面积22200平方米，其中地上建筑面积18550平方米，地下建筑面积3650平方米。建筑檐口高度23.3米，台塔高度30.1米。建筑层数为4层，基本使用功能包括大剧院、群艺馆、电影院、美术书法创作展览馆。剧场按1157个座位规模设计，建筑等级为乙级，使用性质与观演条件为兼顾歌舞类与大型会议类用途。电影院分4个厅，共383个座位，满足市民的多元文化需求。

图 3-3 城市公园方向效果图

图 3-4 西立面效果图

图 3-5 北立面效果图

图 3-6 南立面效果图

图 3-7 东北侧鸟瞰效果图

二、吉安文化艺术中心设计构思及设计过程

（一）城市空间的整合和修正

艺术中心在建筑设计布局阶段需解决两个重要问题，一是与周边城市环境的关系，重点来说就是与西侧吉州大道、东侧市民广场的关系；二是如何创造积极的外部空间。而这两个问题与建筑的功能布局密切相关。芦原义信在其《外部空间设计》中提出："由于外部空间不是无限延伸的自然，而是'没有屋顶的建筑'，所以平面布置是比什么都重要的。"

经过对周边城市环境既有空间和城市轴线的调研分析，艺术中心在基地东侧和西侧分别退线形成东西两个外广场，并在基地中部利用灰空间的处理形成连接东西两个外广场的内广场。通过建筑内外一系列广场的设置，将项目用地东侧的市政府前市民广场引入文化艺术中心建筑内部，营造出连续、丰富的城市公共开放空间。同时文化艺术中心及其内广场，成为联系行政中心前的市民广场与其西侧城市建筑的一个视觉通廊，在市民广场进行活动的人们可以透过艺术中心，直接看到远处城市建筑的对景。通过这一处理，原本空荡无序的市民广场终于在城市尺度上有了方向性和渗透感，并使市民广场东侧的略显单薄的体育馆

（面积较小）成为东西向视觉通廊的一个节点。在这里城市外部空间和建筑外部空间进行了延续和融合，可以说建筑的外部空间处理是十分积极的。

中国传统建筑也十分重视其外部空间和氛围的塑造，中国传统的院落，其“埏埴以为器，当其无，有器之用。凿户牖以为室,当其无,有室之用”的外部空间甚至比内部空间在体验感上要更占据统治力。

反观国内的很多剧场及观演建筑，其建筑造型的处部处理很突出，但内部空间塑造十分无力，致使剧场内部的氛围和活动极其封闭，无法延伸到建筑外部。建筑外部空间不进行界定，十分消极，甚至感觉不到其外部空间的归属感。目前大多数集中式布局的剧场都会有这个缺陷，北京的国家大剧院尤为突出，其周边水系的设置和建筑的完整封闭性，从城市角度都在拒人于千里之外。而这一点上，青岛大剧院就采用分散和开放式的格局塑造了灵动和极具艺术感的剧院外部空间体验。

（二）分散式布局

通过对艺术中心的外部空间的处理和整合，艺术中心建筑体量分为南北两个部分。艺术中心的诸多功能如大剧场、群艺馆、电影院和多功能剧场等均据此进行拆解，形成了分散布局。大剧场及其附属功能位于用地北部（A区），群艺馆位于西南部（B区），电影院及多功能剧场位于东南部（C区）。整体分为南北两大部分，它们之间通过一个“有顶”“无墙”的内广场相互联系。采用这种分散式布局具有以下优点。

①大剧场、电影院、多功能剧场等均为人员密集的场所，分散式布局可有效快速地进行人群的疏散。

②文化艺术中心将多种功能复合于一个建筑，有多个人群流线。分散式布局将人员分流，流线简单明确。

③分散式布局有利于文化艺术中心日后的运营管理，大剧院夜间运营居多，其他功能日间运营居多。

④分散布局最大限度地满足了建筑主体尽可能多的城市界面，确保了建筑主体的自然采光和通风，

图 3-8 东北侧鸟瞰实景

图 3–9 东立面夜景

图 3-10 北立面夜景

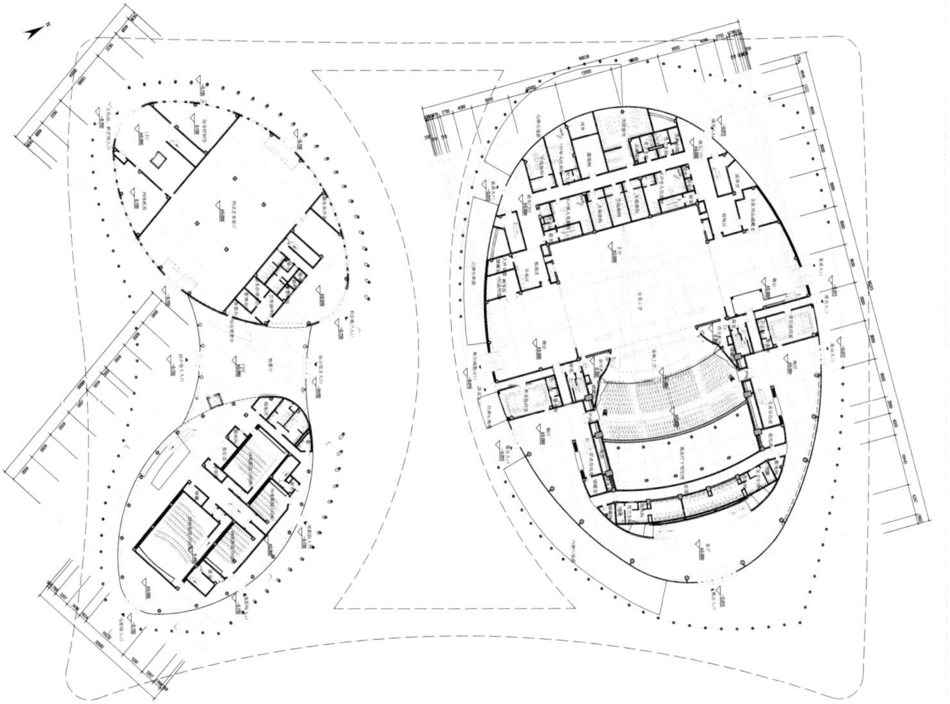

图 3-11 一层平面图

并使得空气在各楼之间形成环流，有利于组织场地通风。借助各个部分的共享空间，形成了理想的气流组织，有助于室内空气质量的提高。

（三）飞扬的屋檐——传统文化的现代阐释

中国剧院的建设一直存在着贪大、媚洋、过度浪费等现象。吉安文化艺术中心的设计力求做到尺度理性，风格清新。江西介于广东、江苏、安徽之间，建筑风格属于南方建筑，偏通透灵动风格。艺术中心在建筑形象的处理上也以此作为借鉴。

建筑南北两部分都采用椭圆形的平面形状。北侧剧场为一个较大的椭圆形。南侧的群艺馆、多功能剧场和电影院为两个较小的椭圆形，这样建筑平面的流线型轮廓喻示了建筑的艺术特性，并在周边创造出流动的外部空间。艺术中心在创作过程中引入了屋顶的概念。屋顶一直都是中国传统建筑的重要且独立的因素，任何中国传统建筑都可以简单地拆分为屋顶和屋顶下部建筑（非民居类可拆分为屋顶、主体和基座三部分）。同时屋顶起翘的处理一直隐喻古代建筑对天宇的承托。

覆盖于南北两组建筑之上的大屋顶，基本呈长方形（长边位于东侧），四周出檐深远，边缘变薄上

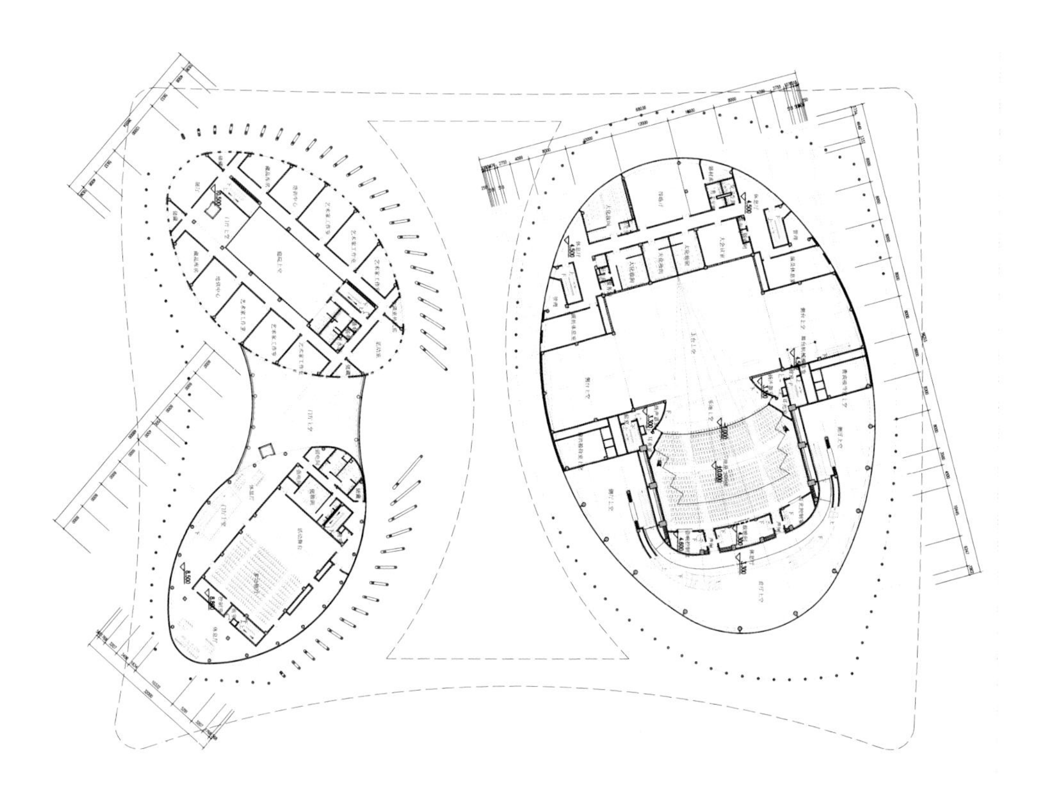

图 3-12 二层平面图

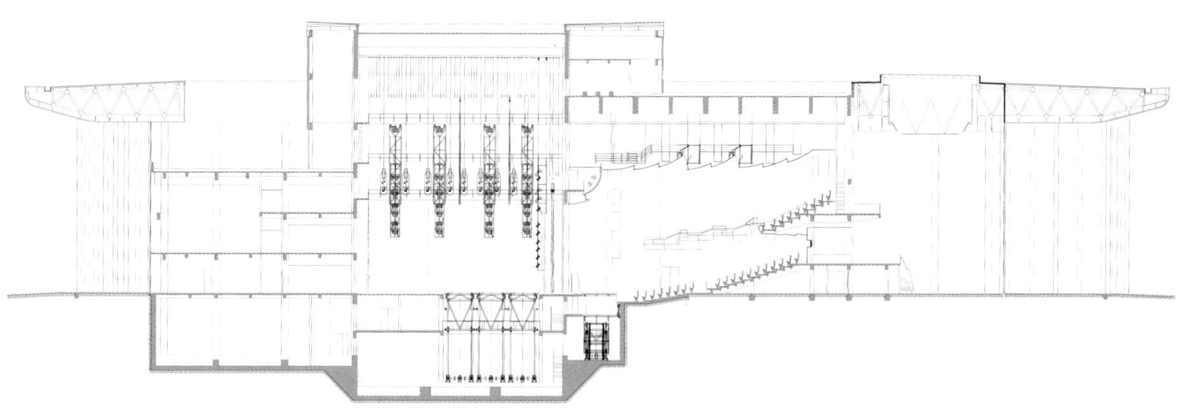

图 3-13 剖面图

图 3-14 外环廊夜景

翘，尤其面向东侧市民广场的屋顶在平面轮廓上处理成为内凹的弧形，这样从透视角度看就形成类似飞檐的外观效果，不但令建筑整体形象具有轻盈质感，也隐喻了古典建筑的精神。更重要的是，屋顶将分散式布局的建筑统一为一个整体，在建筑尺度方面与东侧宽广的市民广场和西侧大尺度的城市主干道相协调，更使通廊形成了镜框式的视觉效果，屋顶最基本的功能（遮风挡雨，限定空间）也在此得以发挥，使内外广场的气候适应性增强并限定了建筑的外部活动区域，一举三得。

目前，对中国古典屋顶的现代诠释多数都采用坡屋面起翘的手法。现代建筑中平屋面较多，因此通过平面的造型处理在透视角度实现古代飞檐的效果，也是一种重要手法。

艺术中心屋顶采用银灰色的铝板饰面，用浅色金属凸显屋顶的轻盈，减少压抑感，同时在内广场的中部进行了单独的处理。结合内广场屋顶结构的网架形式，建筑在这部分设计了透空的树叶造型，使得内广场的屋顶通透轻盈，并隐约折射了镂空窗花和剪纸的艺术特性，阳光透过叶片间隙洒下，体现一种在城市环境中回归自然的情怀，同时树叶的形态也与吉安著名的文物《木叶天目盏》（吉州窑代表性瓷器）相暗合。

（四）灵动的琴弦——艺术的联想

艺术中心另一个重要的造型要素就是屋顶下极具韵律、尺度纤细的立柱。细细的白色立柱密布在建筑实体周围，通过立柱的限定形成了建筑周边带状的灰空间，使建筑外部空间不但在形状和尺度上有变化，在层次上也区分开来。上下铰接的立柱既是承托大屋顶的结构构件之一，同时也是建筑造型的寓意之所在。细柱的排布形成有韵律的节奏变化，形态好似被拨动的琴弦，而轻盈起翘的屋顶又使整个建筑隐喻飞扬之意，令观者产生有关于音乐、艺术的通感式联想。艺术中心的设计案名“弦动庐陵”（古吉安称庐陵）就是据此而来。

图 3-15 杜鹃花

图 3-16 木叶天目盏

图 3-17 中国南方传统建筑，屋檐角部上扬

图 3-18 东南侧鸟瞰图实景

（五）空间渗透，活力社区

艺术中心在设计过程中十分注重内部空间与外部空间及城市空间的交融。南北两组建筑朝向东侧广场侧分别设置了电影院、多功能剧场的休息厅和大剧院的门厅。休息厅和门厅基本在环建筑物椭圆平面的内部周边，这样充分将建筑内部的公共空间贴临建筑表面，同时剧院门厅采用了两层通高的设计，内部空间高耸并富有动感。建筑公共空间的立面材料选定为玻璃幕墙。白天，颜色稍深的玻璃幕墙将屋顶和立柱清晰地映衬出来，突出了建筑的艺术形象，同时远处市民广场及周边城市景象也通过通透的玻璃幕墙纳入门厅和休息厅中；夜晚，通透的玻璃幕墙将建筑内部缤纷的色彩和人的活动展示给市民广场以及整个城市，使建筑的内部生活融入到整个城市生活中去。

而在建筑内部空间渗透下的内广场和外广场，经常结合剧院、影院、群艺馆的功能，定期举行建筑外部的文化艺术活动。这种内外的互动、渗透和融合，使得艺术中心在城市中具有非凡的生命力，地标性的建筑为城市带来地标性的生活。吉安文化艺术中心真正成为了吉安城市生活的观演厅。

艺术中心是公共性极强的建筑，也往往是一个城市地标性的建筑，更是一个城市文化的名片。如何在

图 3-19 以《木叶天目盏》为概念原型的室内中庭

图 3-20 东侧广场夜景

图 3-21 外环廊

图 3-22 外环廊

图 3-23 内庭院

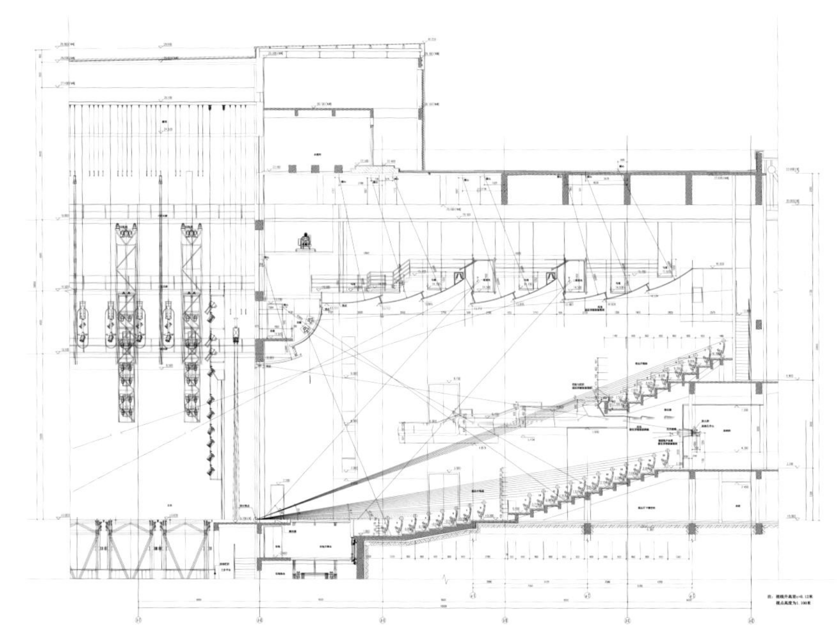

图 3-24 观众厅剖面图

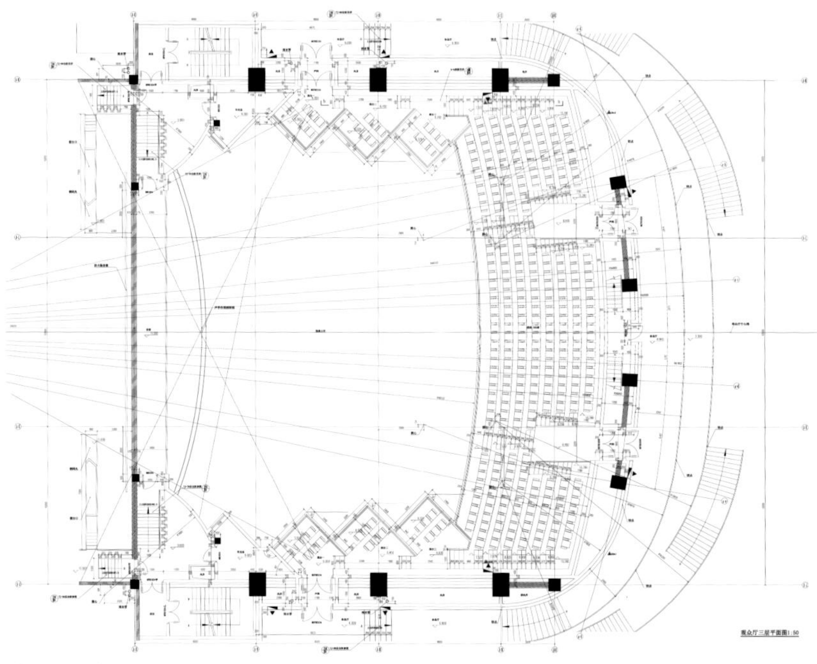

图 3-25 观众厅平面图

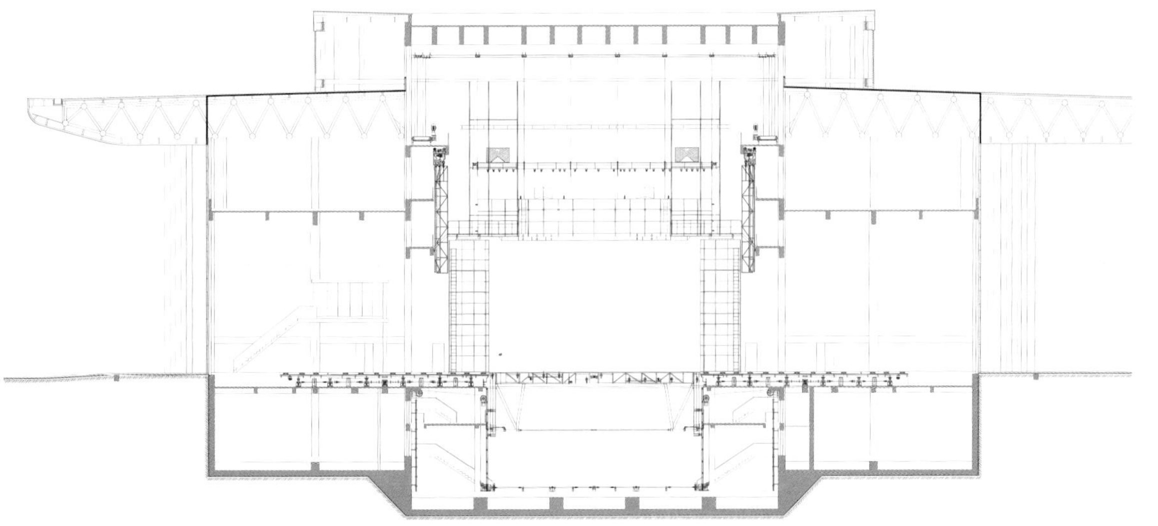

图 3-26 舞台剖面图

细心营造内部殿堂级观演效果的同时，把观演的行为和生活模式带到外部空间，提升周边城市空间的艺术性和参与性，是我们在建筑创作中应该特别注重的。

在中国，当今城市设计在规划中普遍缺失，建筑单体则“各为其政”。在这样的城市环境下，如何将项目周边的城市设计提高到艺术层面，是每个建筑单体设计中都应该给予首要关注的问题。

（六）建筑的城市属性

建筑对自身功能的完美塑造只是建筑实现自身价值的有限意义，而通过建筑艺术化地提升周边城市空间并增添标志性的城市生活，则是建筑对自身价值的升华。生活中绝大多数“可体验”的建筑都坐落于城市中，因此建筑的确具有一个根深蒂固的属性——城市属性，但由于其地理位置、时代背景、功能各不相同，不同的建筑在城市中的作用也千差万别，有的严谨内敛、规矩方整，成为城市肌理中的背景部分，而有的则是城市的重要节点，起到整合城市空间、对周边城市区域画龙点睛的作用。吉安文化艺术中心就是这样一个位于城市重要节点的项目。由于自身独特的功能设定和特殊位置及环境背景，因此在建筑创作方面采用了一些因地制宜的独特设计手法。在这里，建筑内部空间与外部空间互相融合映衬，建筑外部空间与城市空间延续和整合，从而实现了建筑对城市的积极作用，使这座包含剧院的文化艺术中心真正成为城市生活的观演厅，对周边城市空间艺术氛围的塑造和提升做出了极大的贡献。

（七）设计总包服务——精湛技术实现项目整体高完成度

设计总承包，即从设计的最开始，甲方将项目的所有的设计内容全部委托给同一家设计单位来完成，包括规划、建筑、结构、室内、给排水、强电、弱电、暖通、景观、幕墙深化、标识系统等多个子项目。设计总承包具有很明显的优点：既大量节约了造价，又节省了配合的各设计公司之间相互协调的精力与时间，并且责任清晰、明确。

吉安文化艺术中心项目采用的就是设计总包服

务，其设计范围包括总体规划、建筑单体、舞台工艺、电影工艺、室内设计、室外景观等。各专业在设计团队的总体协调下密切配合，高度统一，实现了舞台工艺、电影工艺与建筑的结构、机电密切衔接，室内设计的细部与结构和建筑立面完美结合，景观设计与建筑表皮和谐交接、互为映衬。各专业以精湛技术为核心实现了项目的整体高完成度。

三、吉安文化艺术中心运营及后评估情况

剧院的建设固然不易，但吉安市委市政府充分认识到管理运营好这一文化机构就更加困难。在设计过程中，吉安市委市政府经过慎重研究，决定在建成后将吉安文化艺术中心的剧院部分由北京保利剧院管理有限公司托管，并根据保利剧院管理方的意见和建议对建筑设计进行了修正。项目建成后成立了赖卫东同志牵头的管理委员会。赖卫东同志建设时即作为业主代表全程参与了项目的建设管理，了解项目情况，对于项目的良好运营起了非常关键的作用。首次托管合同期为5年，按照“市场运作、业主监管、委托经营、政府补贴”的经营管理模式，每年吉安市支付托管费用800万元。保利吉安大剧院自2013年1月1日起交付托管，至2017年12月31日5年期满。吉安文广新局针对5年的运行情况进行了后评估。评估情况如下。

（一）托管经营的基本情况

①着力把吉安大剧院打造成吉安人民的高雅艺术殿堂和百姓大舞台。截止到2017年5月31日，吉安市保利大剧院管理有限公司共完成自营演出160场，通过引进国内外的芭蕾舞、交响乐、话剧、音乐剧等高雅艺术演出，让吉安人民不出吉安就能欣赏到国内外的高雅艺术表演，拓宽了人们的文化视野。在安排高雅艺术进剧场的同时，安排本地专业剧团及群众文艺演出201场，丰富了人民群众的文化艺术生活，潜移默化地提升了市民的文明素养。吉安大剧院已成为吉安的一张文化名片，年均81场演出。同时吉安大剧院每年为政府四套班子和指定的全市性重大活动提供服务20场，每年组织开展12场次市民免费开放活动，每年的复合使用天数为113天。

②着力把大剧院打造成展示吉安文化软实力的窗口。吉安保利大剧院管理有限公司以管理有序、服务周到的高标准要求，接待中央及外省市参观考察团225批次，年均51次。考虑到对外演出以晚上为主，利用白天闲时满足吉安城市对外宣传的需要，利用舞台超大LED电子屏幕，播放市情片、专题片和吉安自创的歌曲MV，向参观的客人宣传吉安的“红古绿”文化，展示吉安改革开放以来取得的丰硕成果，为参观的宾客留下了“老区变新区、吉安变化大”的深刻印象，成为兼具对外接待宣传功能的大剧院，是市委市政府对外宣传的一块重要阵地。

③着力加强管理、确保大剧院国有资产保值。吉安保利大剧院管理有限公司建立了一整套人员管理、市场经营、设备定期保养维护、内部财务审计等规章制度。由于管理有序、责任到岗、定时排查、定期保养，大剧院的高压配电、舞台机械、舞台灯光音响、中央空调等重要设备完好率常年达到98％。同时，根据实际操作的需要，公司对部分设施进行了改造升级，确保了国有资产的保值。5年来未发生任何重大安全和设备事故。2013年和2016年连续两次获中国ISO9001质量管理体系认证证书。

④着力维护大局、树立吉安文化艺术中心整体良好形象。吉安大剧院是吉安文化艺术中心的一部分，在管理运营中，吉安保利大剧院管理有限公司根据市委市政府“创建文明城、创建卫生城”等中心工作部署，除做好合同约定范围内的各项管护工作，常年保持大剧院干净整洁的良好形象外，还应甲方的要求，对艺术中心外围的广场、停车场、屋顶防漏排水等进行日常维护，对周边环境卫生进行日常保洁，积极配合市里的重大活动和重要节日开启室外亮化，较好地展示了吉安文化艺术中心的整体形象，为“双创”工作做出了努力和贡献。大剧院连续三年获市级“文明单位”称号，连续两

图 3-27 剧场门厅

年获“吉安市消防工作先进单位”称号。

⑤吉安文化艺术中心先后获得了首都第十七届城市规划及建筑设计方案汇报展优秀方案奖、首届“江西省十佳建筑”称号、第十八届北京市优秀建筑工程设计项目二等奖、全国优秀工程勘察设计行业奖建筑工程二等奖等多项奖励，其典雅的造型设计获得了一致好评。作为吉安城南新区起步区的标志性建筑之一，该项目使吉安城市建设水准、政府管理部门、普通公民的审美意识均得到了快速的提升，新建设项目的风格、材质、色彩等均考虑和文化艺术中心奠定的典雅清新基调一致，使城南新区城市风貌和谐统一。

⑥托管经营中存在的主要问题。一是全年演出剧目分类不准确，部分剧目水平较差，观众反映差，上座率未达到65％。二是政府拨付的资金使用情况，虽然接受了年度财务审计监督，但没有按照合同约定报送财务报表于甲方审核。三是没有区分公益性演出和经营性演出，对外服务收费不合理。四是每月一次的免费开放日活动没有实际内容，流于形式。

（二）后评估建议

①继续委托北京保利剧院管理有限公司托管。目前，北京保利剧院管理有限公司在全国各地托管了

图 3-28 剧场内部

53家大剧院，建立了一整套人员管理、市场经营、设备维护、内部财务审计等规章制度，拥有庞大的剧目资源和专业技术人才队伍，院线管理有序、机制灵活、保障有力、服务规范。继续托管有利于大剧院管理的连续性，保障大剧院正常运转，确保吉安大剧院的国有资产保值不损，有利于引进剧目演出正常进行，丰富市民的文化生活，实现文化惠民的目标。

②续签托管期限为5年，从2018年1月1日起，到2022年12月31日止。每年托管经费1000万元（在目前保利托管的53家大剧院里，该项目的托管经费是最少的）。主要经营指标为每年完成自营演出剧目30场以上，其中A类9场、B类12场、C类9场，上座率达到65％以上；每年免费为市四套班子和甲方指定的全市性重大活动提供20场服务；每年组织开展12场次市民免费开放活动；剧院设备完好率达到98％以上，综合管理获得中国ISO9001质量管理体系认证证书。

四、总结

在我国快速建设剧场的过程中，呈现出以下特点。剧院作为城市文化生活的重要容器，每个城市无疑都希望大剧院能成为其地标，成为城市名片。但在实际运维中，人们往往重视剧场硬件、轻视支撑剧场模式背后软件，按照国外的场团合一制或保留剧目制配置硬件，忽视战略定位、市场规则、商业运营、管理制度建设，造成建成后以收低额的场租为主要收入。另外，重视建设，轻视建成后的运营的现象比比皆是。根据国家大剧院院长陈平的统计数据可以看到，除北京、上海、广州等一线城市外，各城市剧院平均演出场次仅为58场，年演出达50次以上的仅占35%。可以看出，剧场运营情况参差不齐，很多项目盲目上马，未进行较好的前期策划，也未提前选择运营主体，造成建成之后每年演出场次较少，同时舞台机械使用率不高。由于运营成本过高，建得起养不起等情况屡见不鲜。同时，每个城市建设的剧场，需满足各种类型演出使用甚至城市两会的要求，造成剧场的功能综合、日益泛化，专业性不足，不能满足专业演出需要。

基于近年我国大型公共建筑，特别是政府投资的公益项目，缺少前期策划、缺少运营策划，后期缺少运营评估，清华大学庄维敏教授提出了建立“前策划—后评估”的设计方法，力图改善设计程序，构建建筑设计闭环反馈机制。根据国务院2016年的《关于进一步加强城市规划建设管理工作的若干意见》，我国要“建立大型公共建筑建筑后评估制度”。

吉安文化艺术中心项目在运行5年之后进行的使用状况评价证明，吉安文化艺术中心的建设作为一个公益项目取得了较好的社会效益，演出场次远大于同等规模城市的其他剧院，群众反映良好。纵观整个设计和建设、运营过程，有以下4点可供其他剧院建设借鉴。

①近年来，中国的剧院和大型公共建筑的建设出现了贪大、媚洋、求怪、逐奢、趋同等现象，吉安文化艺术中心，力求做到尺度理性，风格清新，形态优雅，汲取了中国传统文化的建筑元素，同时也吸收了西方的现代建筑理念、具通过对吉安本地历史的研磨，塑造出精致典雅，有中国江南文化底蕴的建筑形态。

②大型公共建筑是一个建筑生命体，设计人要树立关注建筑全生命周期的理念，即不仅仅关注建筑的设计和建造过程，更需关注建筑建成之后的运营情况、设备的运转情况、物业管护情况等。

③在设计前期要根据不同城市的经济文化发展情况等提前确定剧院运营主体，运营期成本的测算力求准确。剧院是城市公益事业的一部分，运营方应力求政府合理补贴，但补贴额度应在政府可以承受的范围，以能带来足够的社会效益为主要目标。

④后期建立专业的运营管理团队，进行使用期评价，为以后的建设提供参考，同时有利于改进建筑运营情况。

项目信息及版权所有

吉安文化艺术中心
业主：吉安市文化局
建设地点: 江西省吉安市
建筑设计：北京建院约翰马丁国际建筑设计有限公司
设计指导：吴亭莉
设计团队：1.朱颖 2.邹雪红 3.王鹏 4.朱琳 5.沈桢 6.韩涛 7.葛亚萍 8.苗启松 9.徐斌 10.张胜 11.周宏宇 12.谷凯 13.熊进华 14.姜建中 15.薛磊
基地面积：48800平方米
总建筑面积：22500平方米
项目状态:已建成
设计时间：2009—2011年
建成时间:2012年
获奖情况：
2010年首都第十七届城市规划及建筑设计方案汇报展优秀方案奖
2012年“江西省十佳建筑”称号
2014年度BIAD优秀工程设计奖公共建筑类一等奖
2015年第十八届北京市优秀建筑工程设计项目二等奖
2015年全国优秀工程勘察设计行业奖建筑工程二等奖
摄影/图片版权：陈鹤/北京建院约翰马丁国际建筑设计有限公司
撰文：朱颖 王鹏 赖卫东

肆记 Note Four

打破镜框舞台

Go Beyond Proscenium Stages

“后戏剧时代”“环境戏剧”影响下的观演关系与空间变革

Relations of Performers and Spectators and Changes in Space under the Influence of Environmental Theater in the Post-theater Era

演出形式、观演空间、新技术，
一直是古今中外戏剧艺术的三个"互动因子"。
剧场的存在为观众与演员提供了一种适于演出形式的环境或场所，
演出形式的突破性变化会对观演空间提出新的要求，
新技术的发展会提供更完美的演出空间，会反过来促进演出形式的突破。
这里从戏剧的发展出发，探讨剧场在后戏剧时代的发展，
着重研究了环境戏剧理论带来的剧场设计的更新即观演共享空间的出现。
360度全环绕无限舞台亦为创造之举。

At all times and in all countries, the form of performance, theatrical space and new technology constitute three interactive factors in the art of drama. The existence of theaters provides performers and spectators a place of delivering or viewing performances. The disruptive requirements for the form of performance would also amount to new requirements for theatrical space. The emergence of new technologies would create perfect performance space, which in turn would stimulate the breakthrough in the form of performance. Starting from the evolution of drama, this paper is intended to discuss the development of theaters in the post-theater era, by focusing on the updated theater design brought by the surrounding-oriented theory of drama, i.e. the appearance of performance and viewing shared space. The 360-degree all-around infinite stage is another creative move.

剧场，英文为theatre，其语源出自希腊的动词theasthai，原意为“看”，但在中国将之称为演剧的场所。演出形式、观演空间、建造技术一直是古今中外戏剧艺术的三个“互动因子”，剧场的存在为观众与演员提供了一种适于演出形式的环境或场所，演出形式的突破性改变会对观演空间提出新的要求，新技术的发展会提供更完美的演出空间，会反过来促进演出形式的突破。李道增先生在《西方剧场·剧场史》中写道：“历史上任何一次剧场空间的变革，都是出于戏剧家、导演与戏剧‘观演关系’上的变革。如何处理剧场中观众与演员之间的‘观演关系’始终是戏剧家、导演与建筑师至为关心的问题。”对于剧场建筑而言，用于承载演员的“演”空间和承载观众的“观”空间一直是其建筑空间组成的核心，在几千年的戏剧和剧场历史中，“有一个不变的原则，即是以表演者为中心，观众则属于旁立的地位”。19世纪末开始，先锋戏剧的实验开始对这个“不变”原则进行探索，20世纪60年代出现的环境戏剧理论则对以表演者为中心的戏剧和剧场正式提出了挑战。

一、“环境戏剧”理论在“后戏剧时代”的实践

（一）“环境戏剧”“后戏剧时代”理论提出

在1965年春季号《戏剧评论》中，理查德·谢克纳发表了一篇题为《环境戏剧六原则》的论文，“环境戏剧”（Environmental Theatre）这一术语由此问世，新型的“观”“演”关系开始正式登上戏剧舞台。根据理查德·谢克纳提出的“环境戏剧六原则”：“①戏剧事件是一整套相关的事物；②所有空间都为表演所用；③戏剧事件可以发生在一个完全改变了的空间里或一个‘发现的空间’里；④焦点是灵活的、可变的；⑤所有制作组成部分都叙说它们自己的语言……技术人员应该成为演出的一个创造性的部分，在环境戏剧中一个组成部分不会因为其他的部分而被湮没；⑥文本不是演出作品的出发点，也不是终点，也许没有文字剧本。”

1999年德国人汉斯·蒂斯·雷曼的著作《后戏剧剧场》则第一次对20世纪70年代后的世界戏剧实践作出了统领性的阐述，同时大胆地将世界戏剧的整个发展史概括为三个阶段：①前戏剧剧场时期（古希腊

图 4-1 埃皮达夫罗斯剧场

图 4-2 中国传统戏楼

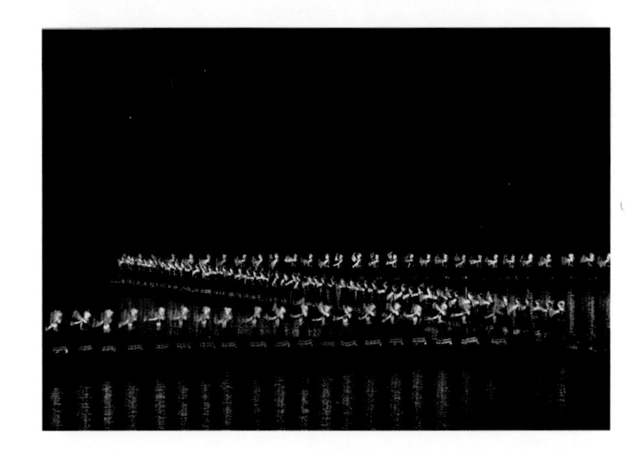

图 4-3 环境戏剧《印象刘三姐》剧照（组图）

戏剧）；②戏剧剧场时期（以剧本为中心的戏剧形式）；③后戏剧剧场时期（萌芽于20世纪初，20世纪70年代开始蓬勃发展，直到今天）。

为了简便，包括安托南·阿尔托（Antonin Artaud）的"残酷戏剧"（Theatre of Cruelty）、波兰戏剧家格洛托夫斯基（Jerzy Grotowski）的"质朴戏剧"、英国戏剧家彼得·布鲁克（Peter Brooke）的"直觉戏剧"（Immediate Theatre）和"浸没式戏剧"在本文中均纳入到"环境戏剧"的理论框架中。

（二）"环境戏剧"实践

1968年开始，理查德·谢克纳通过一系列戏剧实践来践行他的理论，包括由他导演的《69年酒神》(1968)、《麦克白》(1969)、《公社》(1970)等。1991年孟京辉导演将《等待戈多》搬上舞台。孟京辉导演把观众请到台上，演员在台下演，他的黑匣子小剧场打破了舞台和观众席的物理界限，推动了小型室内"环境戏剧"在中国的实践。2003年《印象刘三姐》在阳朔开演，开创了中国"大型山水实景演出"这一新的舞台表演形式。以"印象""山水盛典""千古情"为代表的山水实景演出目前在国内有超过40场，是"环境戏剧"在中国的成功实践与发展。

头晕眼花剧团（Punchdrunk）创作的"浸没式式戏剧"《不眠之夜》（*Sleep No More*）的演出场所是拥有约百间房间的旅馆，不同房间上演不同的故事，观众戴上面具穿梭于不同房间，观众观剧体验独一无二。《不眠之夜》于2016年进入中国。在某种程度上来说，浸没式戏剧是环境戏剧在后现代语境下的一种延伸，浸没式剧场既是一个演员的表演空间，也是一个观众的漫游空间。

近年国内的《又见平遥》《又见五台山》《法门往事》《知音号》和孟京辉执导的《死水边的美人鱼》及赖声川执导的《如梦之梦》引发的对于保利剧院的改造也可以视为"浸没式戏剧"的实践。

图 4-4 浸没式戏剧《不眠之夜》剧照（组图）

二、“环境剧场”“浸没式剧场”等定义与研究现状

（一）“环境剧场”与“浸没式剧场”概念

国内对于为环境戏剧提供演出的场所目前有多种定义，包括“环境剧场”“情景剧场”“情境剧场”“实景剧场”“山水剧场”“浸没式剧场”等，在本文中统一定义为“环境剧场”，包含所有可以为演出提供环境的将“观”“演”空间合一的室内和室外场所。

（二）“环境剧场”观演模式与空间创新

剧场的实质在于演员与观众之间的空间关系。环境戏剧的基本特征是对戏剧空间的创新性使用和观演关系的变革，是一种以对戏剧空间的使用为基础，观演关系中“演”不再是唯一的中心，观众参与到演出中的一种体验式戏剧的变革。

“环境戏剧”的出发点，谢克纳将其描述为“对舞台和观众厅的废止，把所有的障碍清除，代之以一体化的空间”。该理论既是对剧目创作和演员表演的指导，也是对空间选择和使用的“宣言”。

传统的剧场中，无论是开放式还是封闭式，作为“演”的空间的舞台和作为“观”的空间的观众厅都是彼此分离的，由舞美承担着联系舞台空间与戏剧表演抽象时空的使命。观众厅乃至整个剧场建筑本身与其中上演着的戏剧作品没有任何的联系。

在“环境戏剧”（Environmental Theatre）概念提出之前，所谓“戏剧空间”，或者说“剧场”，除镜框式剧场外，不外乎观众从三面或四面包围舞台（伸出式或中心式剧场）。剧场有没有不变的模式？ 它的本质又是什么？一个世纪尤其是最近30多年以来，西方戏剧家思考与探索的正是这些问题，而环境戏剧则是对这些问题的一种回答。

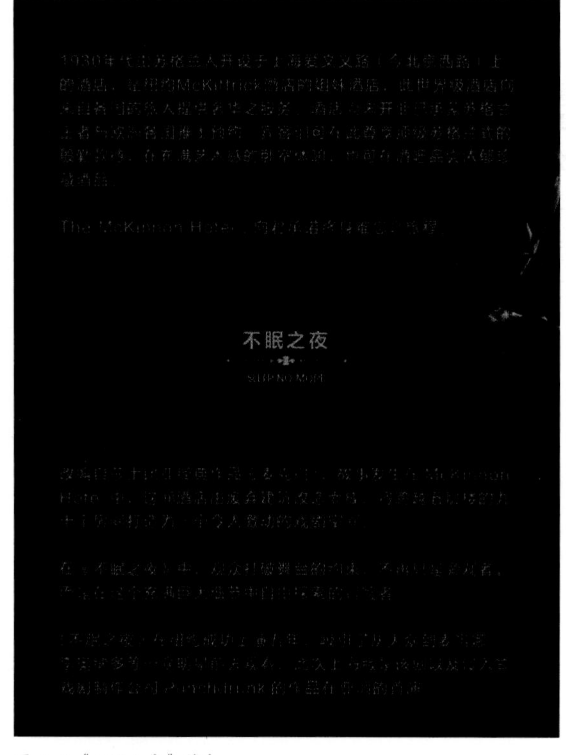

图 4-5《不眠之夜》简介

（三）“环境剧场”国内研究现状

一直以来，对于剧场建筑的研究更多地集中在观演二分的常规剧场空间以及观众厅声学等技术性问题上，环境戏剧作为一种特殊的戏剧革新形式，对观演环境乃至整个剧场建筑都提出了特殊的要求。

在建筑学领域，无论是国内还是国外，对剧场建筑的研究都主要聚焦于传统的歌剧院、音乐厅、戏剧场的分类。李道增先生的《西方戏剧·剧场史》等著作在观演空间上的研究仍大都侧重于传统的“舞台—观众厅”二分的空间。卢向东2009年著的《中国现代剧场的演进——从大舞台到大剧院》也是基于对镜框式舞台包括品字形舞台等传统剧院的研究。魏大中、吴亭莉、项端祈等著的《伸出式舞台剧场设计》对伸出式舞台进行了研究和探讨。

总之，与戏剧学科对空间观念和观演关系的大胆突破相比，建筑设计领域在这方面的研究相对滞后，鲜有从建筑学角度出发的空间层面或观演关系层面的专门性、针对性的研究。著名建筑师朱小地近年完成了一系列环境剧场作品，在建筑设计界甚至引发了关于“还是不是剧场”的争论。根据戏剧界的认识，环境戏剧是戏剧在20世纪最重要的发展之一，无疑环境剧场作为环境戏剧的演出场所，“还是不是剧场”的争论应该可以告一段落。

可变剧场（Flexible Theater）主要是指可以通过机械手段或者人工调节来改变观演关系、舞台形式、声学效果、座位数量以适应不同艺术表演形式要求的剧场。多用途剧场（Multi-purpose Theater）是在技术条件有限的情况下，有多种用途的剧场，常常以牺牲演出效果为代价。

可以看出，“环境剧场”与“多用途剧场”“可变剧场”虽然有时有概念的重合，但有本质的不同。

三、打破镜框式舞台——从镜框式舞台到全环绕舞台

（一）从镜框式舞台到开放式舞台、全环绕舞台

从巴洛克时期至今，历经几百年，镜框式舞台剧场在各种剧场类型中一直占据主导地位。镜框式剧场的基本特征在于画框似的舞台框的存在，它把统一的戏剧空间分成两个彼此隔离的区域：舞台与观众厅，其结果是将戏剧活动的参与者分裂成两个分离的集团。

以观众对舞台（包括室内外舞台）的正面关系为一端，非正面关系为另一端，两者构成舞台围绕观众的连续统一体。正面的一端是像镜框式剧场、中心式剧场和伸出式剧场这类常规的演出，非正面的一端是舞台完全包围观众的演出。两端之间形成一系列中间环节。

本文将专注于室内剧场观演关系的变化，对于室外环境剧场将不做过多陈述。

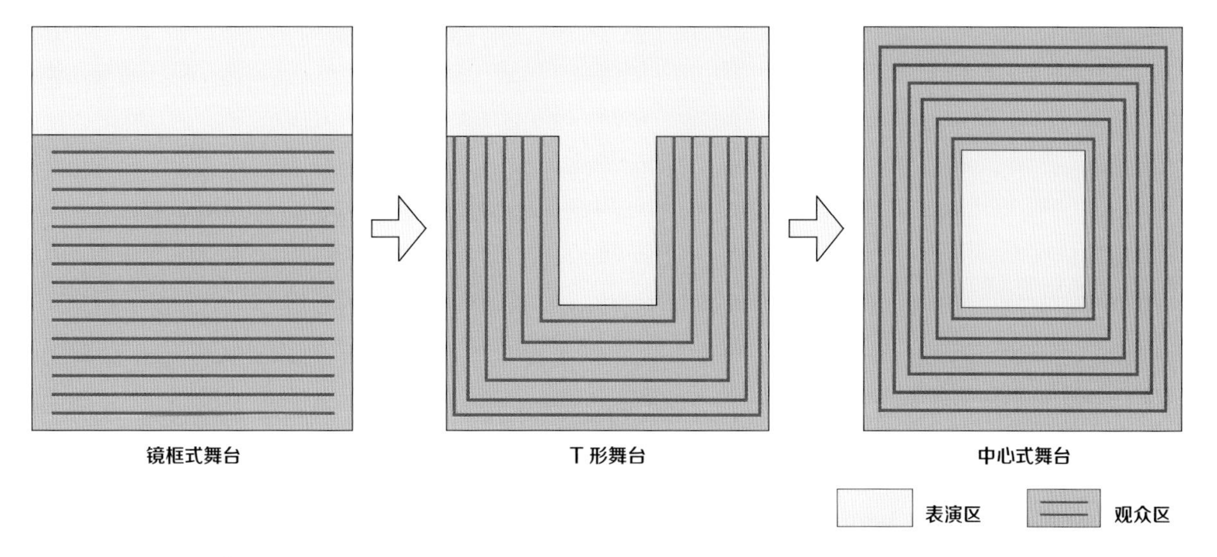

图 4-6 剧场观演关系分析图 1

为了更好地分析室内环境剧场对于观演空间的影响和进化的脉络，本文针对室内剧场的舞台与观众席的关系从“观”“演”二元化到“观”“演”共享空间的进程分为三条主线分析。

①从镜框式舞台到伸出式舞台、T形舞台、中心式舞台的观众席逐渐包围舞台的一条主线。

随着舞台布景的增多，舞台进深的加大，观众和演员之间隔着乐池、台唇、脚光和大幕，观众对于舞台的围合感越来越弱，观演“亲密感”遇到了挑战。“如果没有演员和观众之间感性的、直接的、活生生的交流，戏剧表演便不能存在。”

几十年来，各种探索的共同特点都是要建立更加亲密的观演关系。最初的一步是演员侵入观众厅。最常见的是演员通过观众席中间的走道走下或走上舞台。这样，过道既是观众区又是表演区，成为局部的共享空间。在21世纪初，这种入侵作为打破舞台框（尤其是镜框式台框）的手段是有效的，不过它在今天已失去新奇感。

如果舞台向外伸出形成T形舞台，观众席就开始包围舞台了，如果观众席继续包围舞台直至形成环状，就会形成中心式舞台，这很类似于体育比赛场馆或者马戏空间。

②从镜框式舞台到卡钳式舞台（C、U形舞台）、全环绕式舞台观众席被舞台逐渐二维围合直至三维覆盖的一条主线。

如果反过来舞台在空间上开始包围观众，像钳子似地从两翼包围前座观众，就会形成卡钳式舞台。在上海戏剧学院演出的《哈姆雷特》中，表演区超出镜框台口沿墙向两侧延伸，并略呈包围前排观众的形势，就属于卡钳式舞台（C形）。不过，卡钳式舞台只是使前排观众有被舞台包围的感觉，但对后排观众来说，他们与舞台的关系依旧是正面的。在日本的歌舞伎中，两侧的花道一直可以伸至观众厅的后部，这比C形卡钳式舞台更加增强观众被舞台包围的感觉，形成了U形舞台。尽管这种被包围感还是局部的，但观众不只在心理上，而且在物理上体察到自己处于戏剧环境之中。

如果卡钳式舞台两侧的伸出部分一直延伸到观众席的后部并相交在一起，这就成了环形舞台，我们可以定义为全环绕舞台。这样，舞台完全包围了观众。

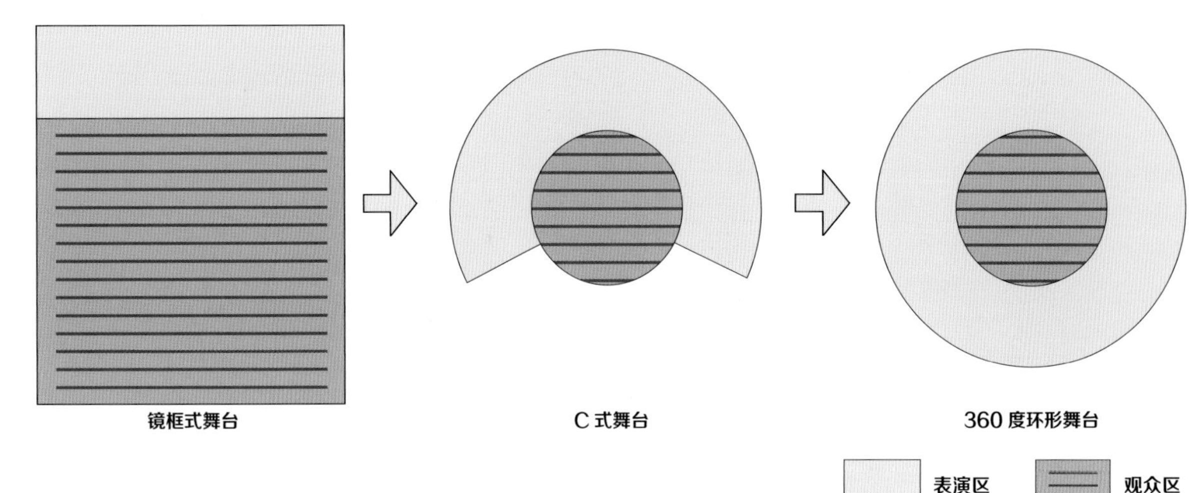

图 4-7 剧场观演关系分析图 2

不过，这样的包围还只限于水平面，如果包围是球形的，观众则处于球体空间的中心。一个世纪以来，许多关于剧场的设计方案追求的正是这种三维覆盖的空间，从戏剧空间的角度来看，这种三维覆盖的剧场无疑是最具有环境性的了。

在20世纪20年代，在包豪斯学院运动主导下，莫尔纳（Molnar）的U形剧院可被认为是现代开放式舞台的概念起源，格罗皮乌斯提出的万能剧院（Total Theater）也是剧院从固定变得灵活的设想 。

③从镜框式舞台到打开台口、“观”“演”部分空间共享，直至全部“观”“演”共享空间的“黑匣子”，到“浸没式剧场”模式。

除此之外，尚有多场所演出、改造场所演出等模式，但每个剧场空间均可在以上三种模式的框架内。观演共享空间在本文第4部分进行探讨。

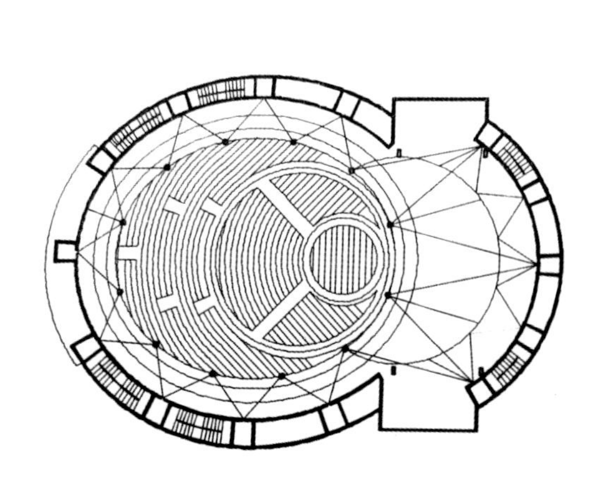

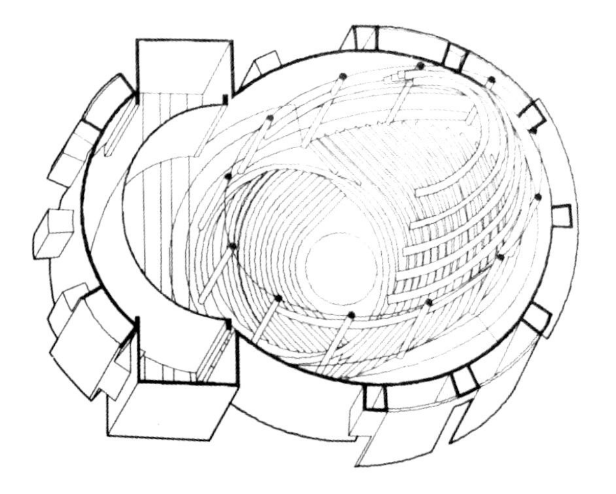

图 4-8 格罗皮乌斯的万能剧院

（二）国内首例全环绕舞台——《又见五台山》专属剧场

该剧场在建筑设计之初，就以确定是为著名导演王潮歌《又见五台山》项目定制设计的。出于大型情景剧演出的需要，《又见五台山》剧场空间是一个长131米、宽75米、高21.5米的大空间。该项目的创作是一次奇妙的体验，导演和建筑师的紧密合作，打破了彼此之间的边界，从建筑主体向前广场延续出连续七折的墙体，宛若徐徐拉开的经折，巧妙地化解了大型建筑体量对自然环境的压迫感，形成没有边际的建筑。在项目设计中，著名建筑师朱小地创造性地提供了一个室内外连续的体验空间，使观众在一个序列中不知不觉地"浸入"到佛教文化里。

《又见五台山》剧场观众人数约1500人。传统剧场通常以后台—舞台—观众席的线性方式布局，台口作为表演区与观演区的分隔将观众的视线局限在镜框型的单向空间内。《又见五台山》剧场以情境体验为核心，创造性地将一个圆形的1500座的可旋转观众席放置在剧场正中心，将舞台表演区围绕观众席布置，使表演场景随观众席旋转能多维度变换，带给观众浸没式的情境体验。360度舞台周长240米，可以根据演出模式调整旋转方向和速度，不到10分钟可旋转一周。根据剧本要求，首演共旋转6周，理论上可以提供总长度近1500米的舞台演出空间，是演出空间的重大突破。

这是我国第一次出现的以表演为主的全环绕舞台，是戏剧和建筑设计的里程碑事件。这一设计首先提供了全程不间断的表演，理论上舞台无限长。其次是时间和运动作为表演的要素正式进入到戏剧的表达方式中。这一项目给戏剧的表达方式带来什么样的影响尚未可知，但作为我国第一个全环绕剧场，带来的感官体验无疑是震撼的。

环绕式舞台最早为基斯勒（Frederick J. kiesler）于1923年设计的无端头方案，但只是作为设计概念提

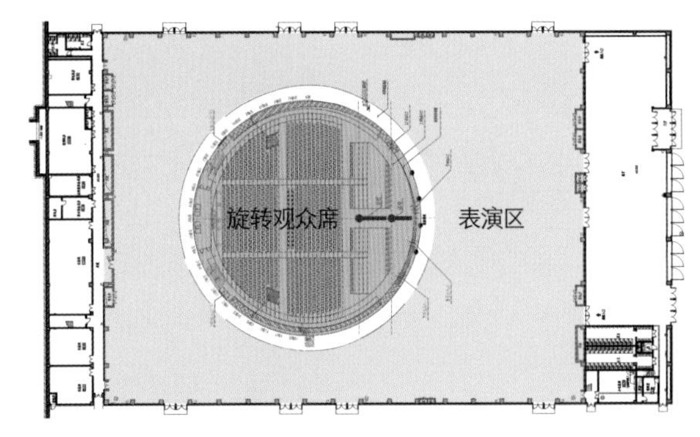

图 4-9 全环绕舞台示意图

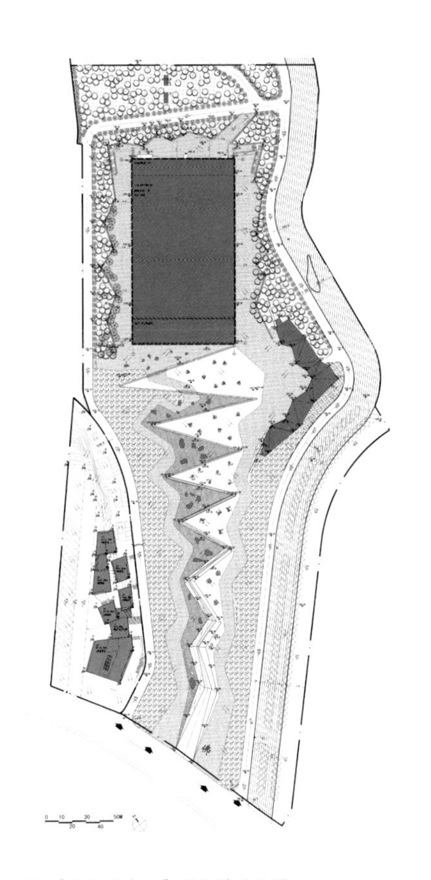
图 4-10《又见五台山》剧场总平面图

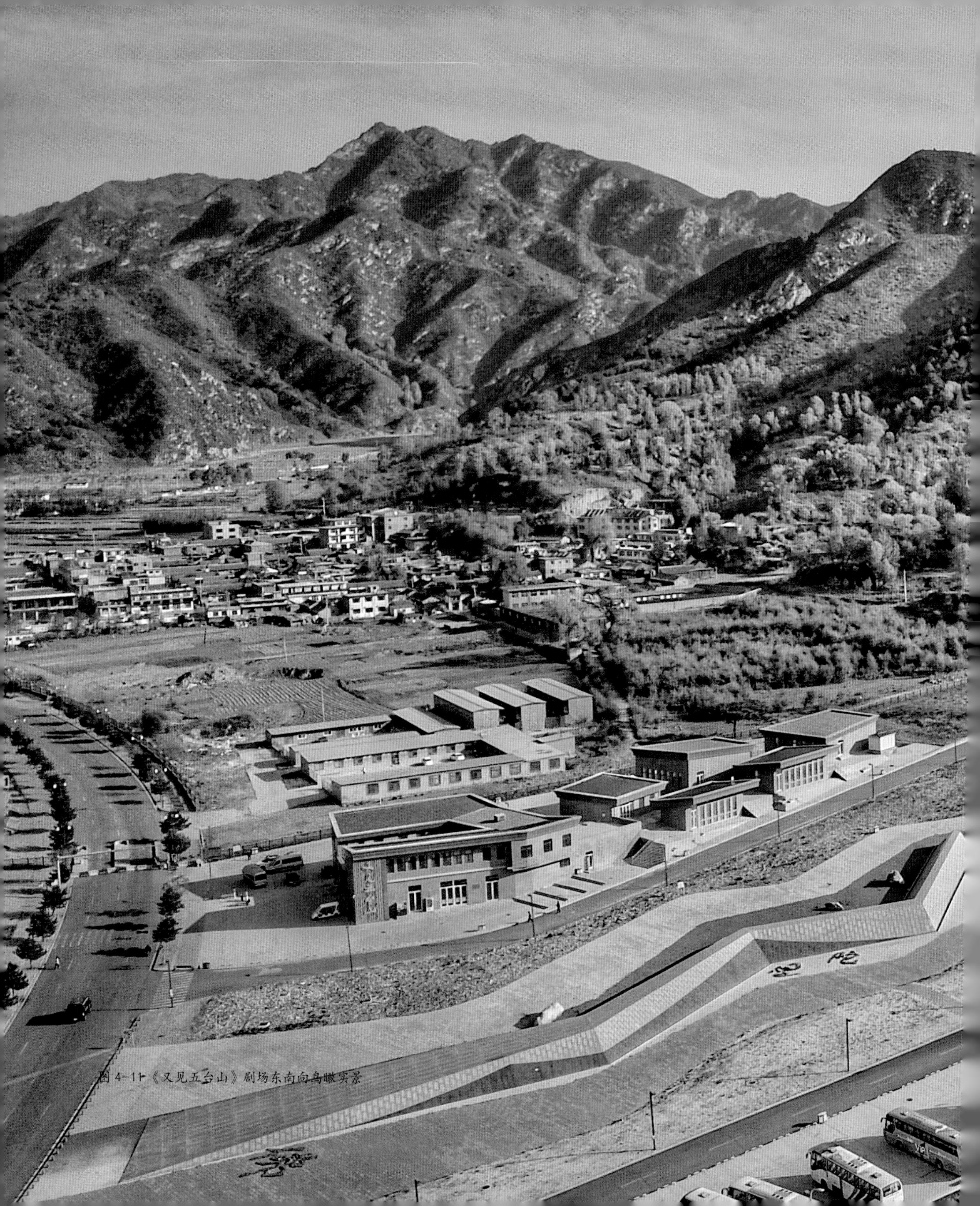

图 4-11《又见五台山》剧场东南向鸟瞰实景

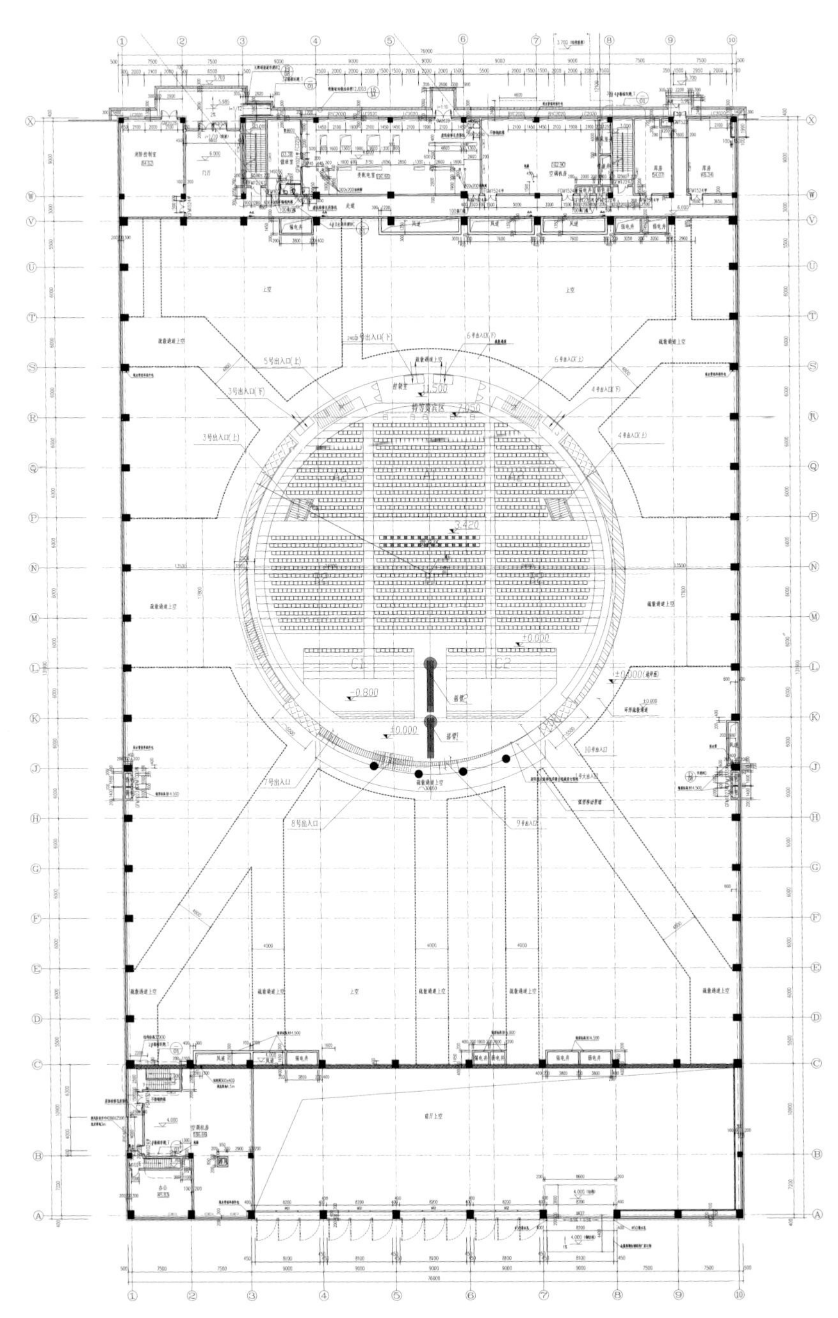

图 4-12 二层平面图

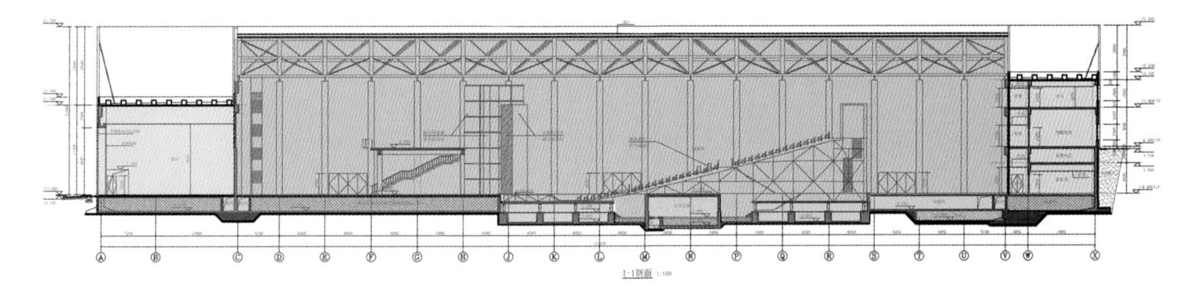

图 4-13 剖面图

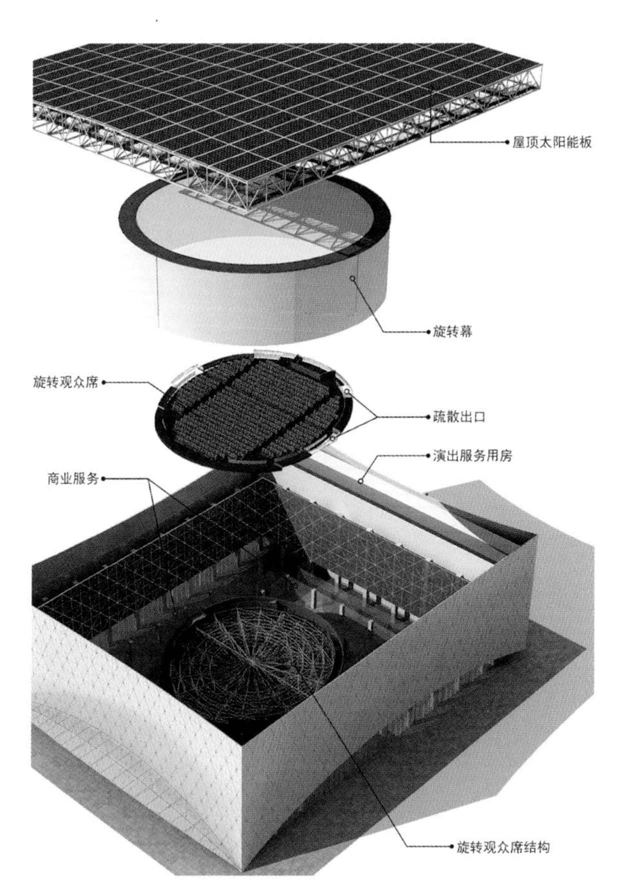

图 4-14 现代工业遗产园区中某 360 剧场设计方案

出，并未获得实施。直到1968年，法国由沃庚斯基设计的600座的蛋形全环绕舞台——格勒诺布尔文化之家（Maison de La Culture，Grenoble）被认为是世界上第一次全环绕舞台360度旋转观众席实践。50年来，全世界的实践寥寥无几，可以查询的仅仅包括2010年荷兰建成的Theater Hangaar，内部装置了360度旋转观众席和3米宽的舞台及环幕影院。可以看出，本项目在观众席的规模、演出场地的空间尺寸、实施的技术难度都远大于仅有的两个全环绕舞台实施的项目。

在《又见五台山》开幕之后的2017年3月30日，日本首座观众席360度旋转剧场 IHI Stage Around Tokyo在东京开幕，国内的橙天嘉禾同样在做全环绕舞台的尝试。

四、观演共享空间类型探讨

本文第三部分的研究均是建立在观众和舞台空间关系的基础上的。在这方面，不管观众受舞台包围的程度如何，他们还始终处于舞台空间之外的一个具有明确界限的区域内。然而，演员可以打破舞台框架，与观众共享表演空间。由此，我们的论述开始转移到观众和演员共享整个戏剧空间的观演空间研究上。

有人曾经认为“剧场就是两块板和热情”，

图 4-15 室内 360 度旋转观众席实景

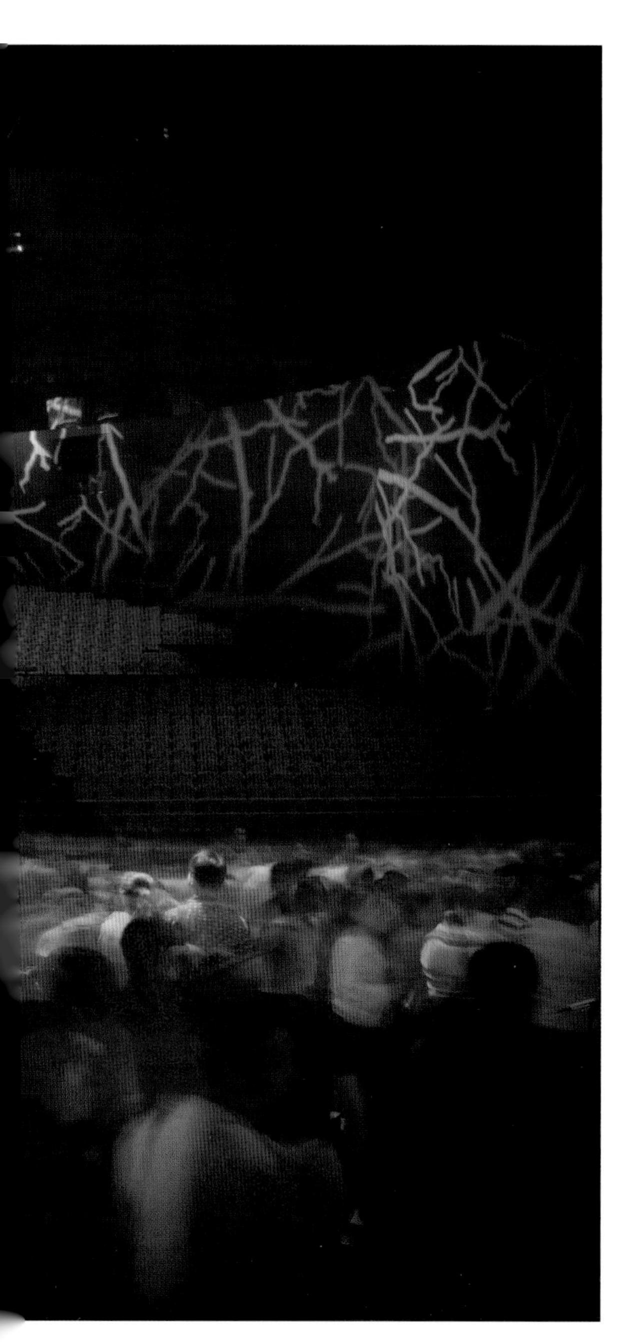

在环境戏剧和沉浸戏剧快速发展的今天，也许应该变成"一个空间和观众与演员之间的热情"更为恰当。部分学者把观众和演员之间、观众相互之间建立一种亲密无间的关系称为剧场空间的"亲密感"（Intimacy），认为这是剧院空间的魔力，是激发观众和演员热情的源泉。观众的体验与参与变得异常的重要，观众不但要实现空间浸入、身体浸入，更重要的是心理浸入。

根据对于现有的环境戏剧的研究，共享空间可以分为以下几种类型："黑匣子"小剧场、共享单厅式、单线索连续多空间式、多线索多空间式和其他定制式混合型及改造型观演空间等，下面分别进行分析。

（一）"黑匣子"小剧场

小剧场打开了镜框式舞台的台口，实现了观演之间的亲密关系，观众视距有利于观众和演员的视线和心理交流。在小剧场中，舞台布置越来越灵活，小"黑匣子"剧场迎刃而生。

黑匣子剧场是一种建筑形式较具有约束性的实验性剧场。黑匣子剧场是安东尼·阿尔托提出的概念，并没有人确切地下过一个严格的定义，但都具有统一的特性（约束性）：黑色或者深灰色的墙壁、可改变剧场内观演关系、演出的戏剧具有先锋性、剧场规模比较小（300人左右）。可以认为，"黑匣子"小剧场是共享单厅式剧场的雏形。

"黑匣子"小剧场是应艺术创作的先锋性和实验性而产生的，需要根据不同的演出形式而改变观演关系，可以适应传统的镜框式舞台、伸出式舞台、中央式舞台、周边式舞台等4种常见模式的观演关系，也可以适应一些以这4种模式为基础的有少量变化的观演模式，或是适应特殊演出安排的特定剧场模式。因此黑匣子剧场内没有特定的观演空间模式，其灵活多变性、多功能性都为先锋性、实验性演出提供更多的可能性。

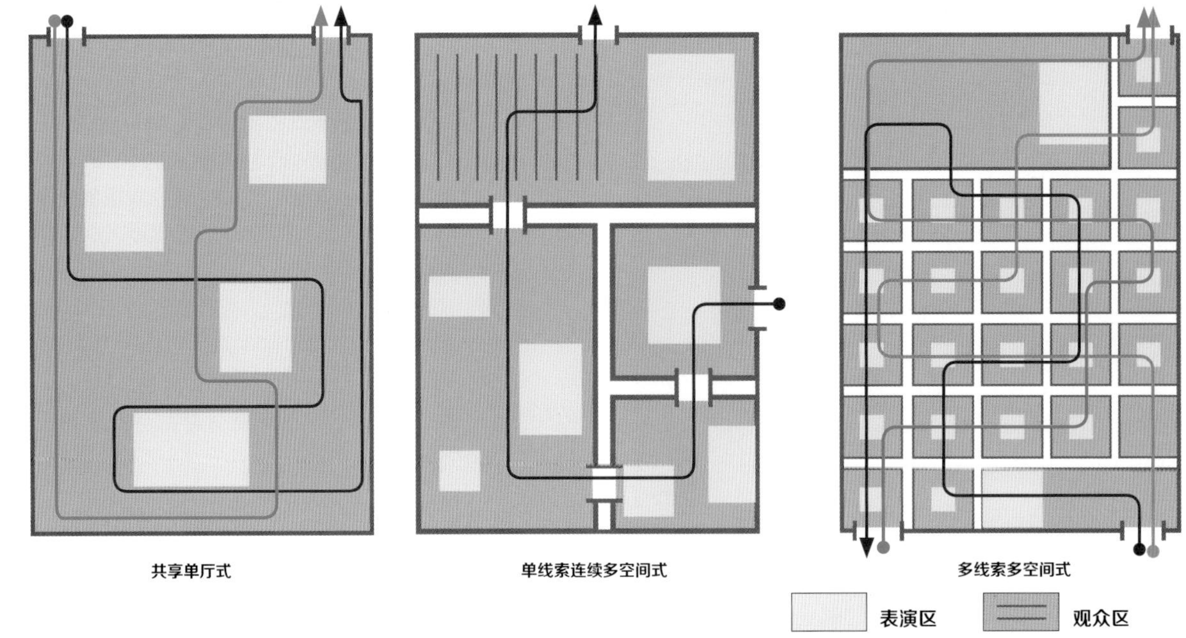

图 4-16 剧场观演关系分析图 3

（二）共享单厅式——散点式舞台、多焦点舞台

"环境剧场"鼓励在一个球形组织的空间里给予和接受——普通空间成为幻觉：观众占据的区域是一个大海，演员在里面游泳，表演区域是观众中间的一些岛屿。

如果表演区置于观众空间的若干点上，这就是散点式演出（Seattorstaging）。在谢克纳的《罪恶的牙齿》中，观众可以在空间内的任何地方随意走动。不过，其中有些区域——某些平台与桥——要用作地点化的空间。一个平台变成一间卧室，另一个平台成为厨房。如果观众已占领表演者所需的任何区域时，表演者就请观众让开。最后，当观众区与表演区之间的界限变得难以划定，或者说，这两个空间完全叠合，形成更大的共享空间时，观众就难以与演出建立正面的关系。在真正的共享空间中没有预定的表演区与观众区，表演者通过他们的行动在观众中建立临时的表演空间。共享空间比非共享空间更具有环境特征，因为它使观众处于戏剧行动的中心。运动与共时性是共享空间的两大要素。运动是指观众可以跟随表演到处走动，这可以影响观众对空间的知觉。共时性表演使观众不能从正面看到整个演出，从而增强了环境特征。

（三）单线索连续多空间式——以《又见平遥》剧场为例

《又见平遥》剧场设计灵感来源于古城内绵延起伏的屋面，有高潮、有低谷，一波三折，起起落落，仿佛在演绎着一幕幕人生大戏，也展示着古城曾经的繁华与沧桑。《又见平遥》剧场的整体思路为具有典型的北方建筑特色的"沙瓦建筑"，剧场内共分了A、B、C、D四个实景演艺区，全部为仿古建筑。A区为引导区，B区为商业文化展示区，C区为生活场景展示区，D区为综合文化演艺区。

《又见平遥》空间类型为典型的单线索连续多空间式。

（四）多线索多空间式——"浸入观展"式专属剧场

《不眠之夜》的演出场所是拥有约百间房间的旅馆，不同房间上演不同的故事，观众戴上面具穿梭于不同房间。"戏"需要观众去寻找，观众观剧体验独一无二。《不眠之夜》2016年推出了具有本土特色的中国上海版。

2017年4月26日，由著名导演樊跃导演的《知音号》在武汉首演，演出的场所是长江上第一艘按20世纪30年代风格打造的大型主题演艺轮船，分上下3层，长120米、宽22米、高15米，98个舱室，整艘江轮都是演出空间，并分为多个区域。所有区域的演出同时开始，船客们分3条不同的线路登船，零距离观看每一个演区的展演。这里没有舞台，更没有传统观众席。知音号的起航，可以视为环境戏剧在中国的新的"漂移式"实践。

《不眠之夜》《知音号》的空间类型为典型的多线索多空间式，是环境戏剧最为复杂的空间形式之一。

（五）其他定制式混合型及改造型观演空间

关于"环境戏剧"的空间设计，理查德·谢克纳在"环境设计六原则"中提出"为每出戏设计整个空间"，即每一出戏设计一个剧场或者说根据每出戏的独特构思设计一个戏剧活动的环境。

所有演出空间和建筑设计、空间设计、舞美设计、舞台机械等，一切均以导演剧目的创作为核心，专剧专演，此类"专属化"观演空间成为观演建筑未来发展的趋势之一。

演出可出现于一个现成的空间内或一个经过改造的空间内。最近30年的一个明显的倾向是从现实环境中寻找戏剧空间，它们可能是现成的，也可以加以适当的改变，包括利用非剧场的建筑物，如仓库、地下室、教堂、菜市场等，利用室外环境打造街头演出、多焦点剧场、在一个空间内布置几个舞台同时进行表演、移动的剧场等。

改造型观演空间也是一个预先存在的空间，不过是根据演出的构思进行了改造。这样的改造是临时的，主要通过布景因素、舞台区域与观众席单元的布置来完成。

五、总结与讨论

（一）加入时间维度的"观""演"空间观

本文讨论的建筑空间一直是基于三维空间的，但在现代戏剧（包括环境戏剧）实践中，我们可以清晰地感受到时间作为一个维度参与到戏剧空间的塑造中来，参与到戏剧创新中来。我们可以认为戏剧空间在从传统的三维物理空间向增加了时间维度的四维戏剧空间发展。例如在《又见五台山》剧场中，同样的舞台演出，观众席旋转的速度、旋转的方式的不同，会造成观众朝向、位置等的不同，人们看到的舞台演出会有巨大的差异。即使是同样的舞台空间，也会有不同的空间感受，比如，同样的演出，顺时针旋转观众席和反时针旋转观众席会得到完全相反的视觉结果。再如，《不眠之夜》的演出中，观众到达一个场所的时间会影响到他对戏剧的认识，几条不同的主线，会得到不同的结果。我把这个视为增加了时间维度的四维空间，这个定义和物理学的定义并不是一样的，但也许这是个更容易理解戏剧空间发展的定义。我们也可以理解为观众的移动也包括观众席的移动会形成时间维度，单个观众的无规律的自行选择的移动会使每个观众得到不同的结果，观众席的整体移动是使其得到一致的时间维度影响下的四维空间而已。无论如何，观众移动使时间作为一个维度存在也许是戏剧演出方式的重要突破之一。至于时间维度对于戏剧空间的影响，需结合物理学、几何学研究方可，故本文仅提出但未能作出深入有效的研究。

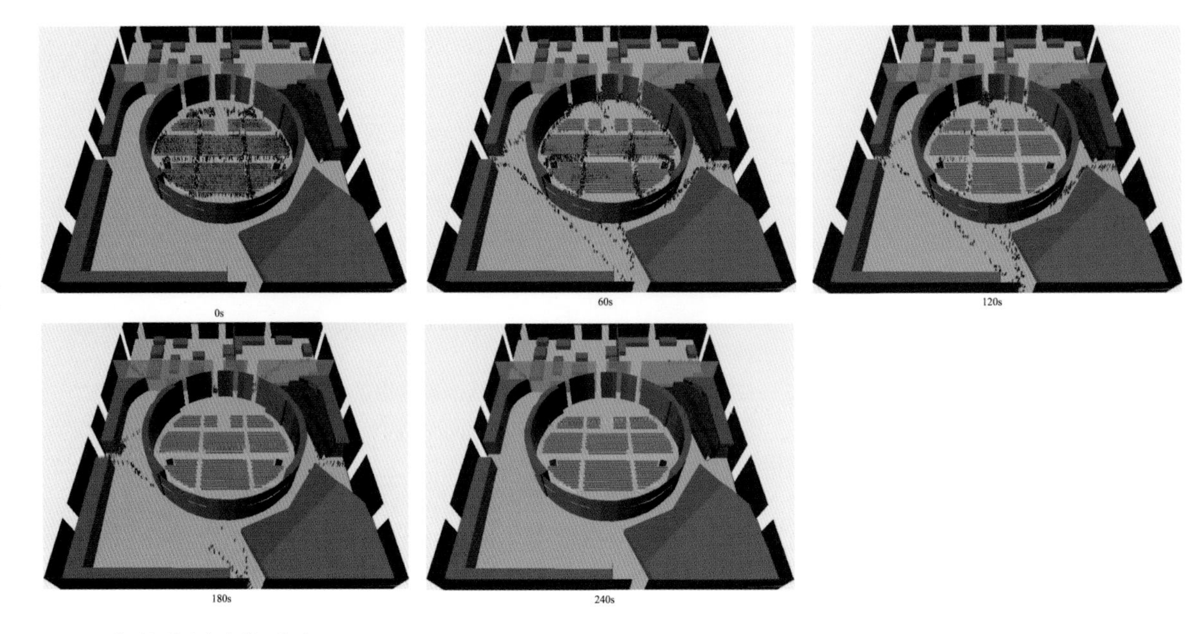

图 4-17 典型场景疏散计算示意图

（二）新型观演场所适用法规缺位及消防疏散等安全问题的提出

演艺建筑作为技术难度相对复杂，创新要求高的建筑设计类型，法律法规的发展严重滞后于戏剧的发展。目前《剧场建筑设计规范》（JGJ—2000，J—2001）的编制基础是镜框式舞台，以观演关系二元化为基础。环境剧场和浸没式剧场在中国的发展显然引发了新课题，一是观演合一型观演空间的设计依据是什么，是作为剧场还是其他的建筑空间？浸没式剧场如果模拟酒店的空间，那设计是按照酒店的要求还是按照剧场的要求？

在《又见五台山》剧场的设计过程中，经历了两轮消防性能化设计，才最终确定了消防设计原则。目前，我国观演建筑防火设计主要设计依据是《建筑设计防火规范》。该规范严格规定了观演建筑的防火分区面积、疏散通道宽度、防烟分区面积、建筑外部扑救条件，等等。这些严格的规定极大地限制了建筑设计中使用空间的尺寸高度以及跨度等。在建筑设计已经成为一种艺术，追求空间形式美的现代公共建筑设计潮流中，现代观演建筑空间尺度上的需求很难满足严格的防火规范，因此，完全根据条文规范进行防火设计在一定程度上严重阻碍了大型观演建筑的艺术创意以及空间设计灵感。

按照传统镜框式剧场的疏散模式，舞台区域与观众席之间有防火幕分隔，两者疏散路线分离，避免交叉。但《又见五台山》剧场的安全疏散面临这样的难题：1500人的观众席疏散必须要经过表演区域，表演区域因舞美设计的需要设有表演道具，疏散路线的迂回、道具的火灾隐患及与演员的流线交叉增添了疏散难度。其次，表演区因为无法与旋转观众席分隔，只能划分为一个防火分区，面积达7737平方米，超出了《建筑设计防火规范》有关防火分区的规定。而作为国内第一个浸没式观演场所，360度舞台包围的旋转观众席为国内首例，设计无先例可循，动态的观众席的疏散设计也是面临的难题之一。

作为专属定制式剧场，舞美设计和演出方式与建设同时进行，导致消防设计中的许多参数和设计依据不能提供，同时旅游演出市场竞争激烈，必须有开创性的令人难忘的演出，才能取得较好的社会效益。创造性地解决问题，并为舞美设计和演出方式预留出条件是设计面临的挑战。

随着新建建筑物的使用功能和性能的发展，它们的防火安全设计可能超出防火设计规范规定的适用范围，或者涉及现行规范未涵盖的内容。由于"环境剧场"设计的特殊性，观演共享空间设计、剧场舞美布景布置方式、使用材料等因素对剧场消防安全、人员疏散等有很大影响，会给演出的安全和疏散包括管理带来很大风险，目前均按照消防性能化设计安排解决方案，但在演出排练过程中，演出方式、舞美设计、疏散设计可能会修改却没有进行消防和疏散的重新论证，可能会造成更大的风险。

（三）戏剧面临的挑战与戏剧空间的未来

社会的快速发展，人的接受方式由"线性、连续性"向"共时性、多层次"发展，旧有的接受方式的原始形态即文学性的文本阅读，已让位于一种更表面化，同时也更全面的接受方式。在相对便捷的影像面前，速度迟缓的阅读和观看失去了优势，以"文本为中心"的复杂和低速的剧场艺术受到快餐式的大众传媒的严重冲击。被业内称为"大导"的林兆华曾表示，"谁能改变戏剧形式，谁就能超越前人"。各种实验戏剧的产生无疑是戏剧面对新媒体时代的竞争做出的尝试和对戏剧发展的探讨。

现代声学技术、多媒体影像技术的发展，给一个"黑盒子"空间的环境创造能力提供了更多可能。在媒介变革时期，浸没式剧场是各门类艺术交叠所指向的一种综合艺术形式，它融合了表演、观念、剧场、舞美、图像、声音、影像、现代APP等多种艺术门类与表演手段，浸没式剧场旨在解放被动观看的观众，不仅仅是身体进入演出空间中，更是充分调动观者的感官并多维度参与到剧目作品的创作之中。

21世纪的今天，中国已经涌现出了新一代戏剧和剧场创作者，无论如何，我们不能对当代创作中的剧场空间作一个定论。戏剧一直向前发展，虽然无法预知在以后的戏剧发展道路上，剧场空间会有什么样的实验和探索，但历史的车轮滚滚向前，一切必将到来。

项目信息及版权所有

《又见五台山》剧院
业主：五台山管理局
建设地点：山西省忻州市
建筑设计：北京市建筑设计研究院有限公司+北京建院约翰马丁国际建筑设计有限公司
联合设计：BIAD艺术中心、BIAD第七设计院、BIAD照明工作室
主创设计师：朱小地
设计团队：1.朱小地 2.高博 3.朱颖 4.罗文 5.田立宗 6.孔繁锦 7.贾琦 8.韩涛 9.田玉香 10.赵伟 11.赵阳 12.王越 13.张胜 14.章伟 15.江雅卉 16.李志鹏 等
消防性能化设计：清华大学公共安全研究院
基地面积：152909平方米
总建筑面积：13949.7平方米
项目状态:已建成
设计时间：2013年
建成时间：2015年
获奖情况：
2015年度BIAD优秀工程设计奖公共建筑类一等奖
2016年亚洲建筑师协会公共建筑设计提名奖
2016年中国建筑学会建筑创作奖金奖（公共建筑类）
2017年中国建筑学会中国建筑设计奖（建筑创作）
2017年北京市优秀建筑工程设计项目一等奖
2017年全国优秀工程勘察设计行业奖一等奖
摄影/图片版权：傅兴 等/北京市建筑设计研究院有限公司/北京建院约翰马丁国际建筑设计有限公司
撰文：朱颖

伍记 Note Five

文化引导设计
Culture-oriented Design

基于运河“源”文化传承、构建文化—山水城市的北京昌平
“白浮泉、化庄、龙山、凤山”城市设计实践

Practices of Urban Design at Baifuquan, Huazhuang, Longshan and Fengshan at Changping, Beijing
Featuring the Cultural Inheritance of the Grand Canal Source and Cultural Construction

城市的文化不仅是城市竞争力的重要标志，更是城市转型最需要的复兴要素之一。
本记基于研究了白浮泉这个昌平的核心历史文化遗存，
探讨了它在北京历史上的生命之根，
进而提出白浮泉乃世界文化遗产中国大运河的重要“点段”之一，是大运河之“源”，
其价值不仅指导该区域城市设计，
也佐证了它在大运河线性遗产中不可或缺的历史地位。
进而深化研究龙山、凤山等诸多历史遗迹，在永安城湮灭的历史遗憾下，
提出一系列重新构建昌平城区文化城市、山水城市的设计手段。

For a city, culture not just serves as one important hallmark of its competitiveness, but comes as a much-needed impetus for its transformation and revival. Based on the findings of the research into Baifuquan, a core historical and cultural site at Changping, this paper discusses the fundamental role the site had played in the history of Beijing and on the basis, concludes that Baifuquan comes as one of the major segments and even the headstream of the Grand Canal, a UNESCO World Heritage site. This paper matters, because it has provided inspirations for the urban design program and also attests to the fact that Baifuquan is an indispensable part of the Grand Canal's linear heritages. Inspired by the findings, the paper goes deeper to study many historical sites ranging from Longshan to Fengshan. Feeling sorry for the disappearance of Yong'an Town, the author puts forwards a series of design skills in an attempt to build Changping a cultural and scenic town.

一、文化侧重型城市设计理论的提出

改革开放以来，我国进入一个快速城市化的历史时期。在快速发展的过程中，大量的文物历史建筑遭到毁灭性的破坏。进入21世纪以来，从文化的意义上来思考城市建设的目标，从文化资源的角度来把握城市发展的可持续性，已经成为全球城市发展的一个重要潮流。单霁翔也提出，以工业革命为背景产生的功能城市，必将走向以文化作为城市发展核心价值的文化城市。

（一）从功能空间引导型到文化侧重型城市设计

城市设计又称“都市设计”（Urban Design），《不列颠百科全书》对其的定义具有代表性：“城市设计是指为达到人类的社会、经济、审美或者技术等目标而在形体方面所做的构思。”城市设计早期的主要特征是信奉“物质空间决定论”(Physical Determinism)，对较大版图范围内的城市形态进行三度形体控制。

《中国大百科全书（建筑·城市规划·园林卷）》中对城市设计做出如下表述：“城市设计以城镇发展建设中空间组织和优化为目的，运用跨学科的途径，对包括人、自然和社会因素在内的城市形体环境对象所进行的研究和设计。”

在近年的城市设计实践中，一个明显的变化是，城市设计从“功能空间”主导型向“文化侧重”型转变。这种转变基于大规模新区建设的减少，基于城市文化遗产的关注和文化城市建设的需要。千城一面引发的城市特色争论与城市记忆湮灭也是城市设计关注文化要素的重要诱因之一。2017年实施的《城市设计管理办法》也提及了城市文化建设在城市设计中的重要性，要“体现地域特征、民族特色和时代风貌”。

我国自古城市设计理论和实践的重要传统是营城从大范围的自然环境出发，结合中国传统士大夫的“文化山水观”，进行人居环境的选址和城市的营造，与西方全盘引入的城市设计有所不同。通过将人居环境置于大尺度的自然山水之中，用文化山水观念进行山水城的总体布局，并在山水的关键节点设置建筑物，营造人与天地、人与山水的宏阔秩序，例如古城西安与终南山的关系、北京旧城的空间秩序，都是中国传统山水与文化观的城市设计实践。

我国的城市设计实践不完全是在物质空间规划的基础上前行，王建国院士在2016年提出的新的城市设计定义：“城市设计主要研究城市空间形态的建构机理和场所营造，是对包括人、自然、社会、文化、空间形态等因素在内的城市人居环境所进行的设计研究和工程实践活动。”可以看出，最重要的改变是在城市设计中增加了侧重了文化要素的引导。

二、北京昌平中心城区形成及城市特色消失

（一）北京昌平区的历史与文化

昌平区位于北京西北部，距离北京市区33千米，素有“京师之枕”的美称，1999年升级为区，区政府所在的原昌平镇本身作为北京的卫星城之一，具有独立的城市属性。为避免混淆，在本文中将昌平区政府所在地具有独立城市属性的原昌平镇称为“昌平中心城区”。

早在6000多年前，昌平境内即有史前人类繁衍

图5-1 郭守敬像（左）及永安城督造人昌平侯杨洪像（右）

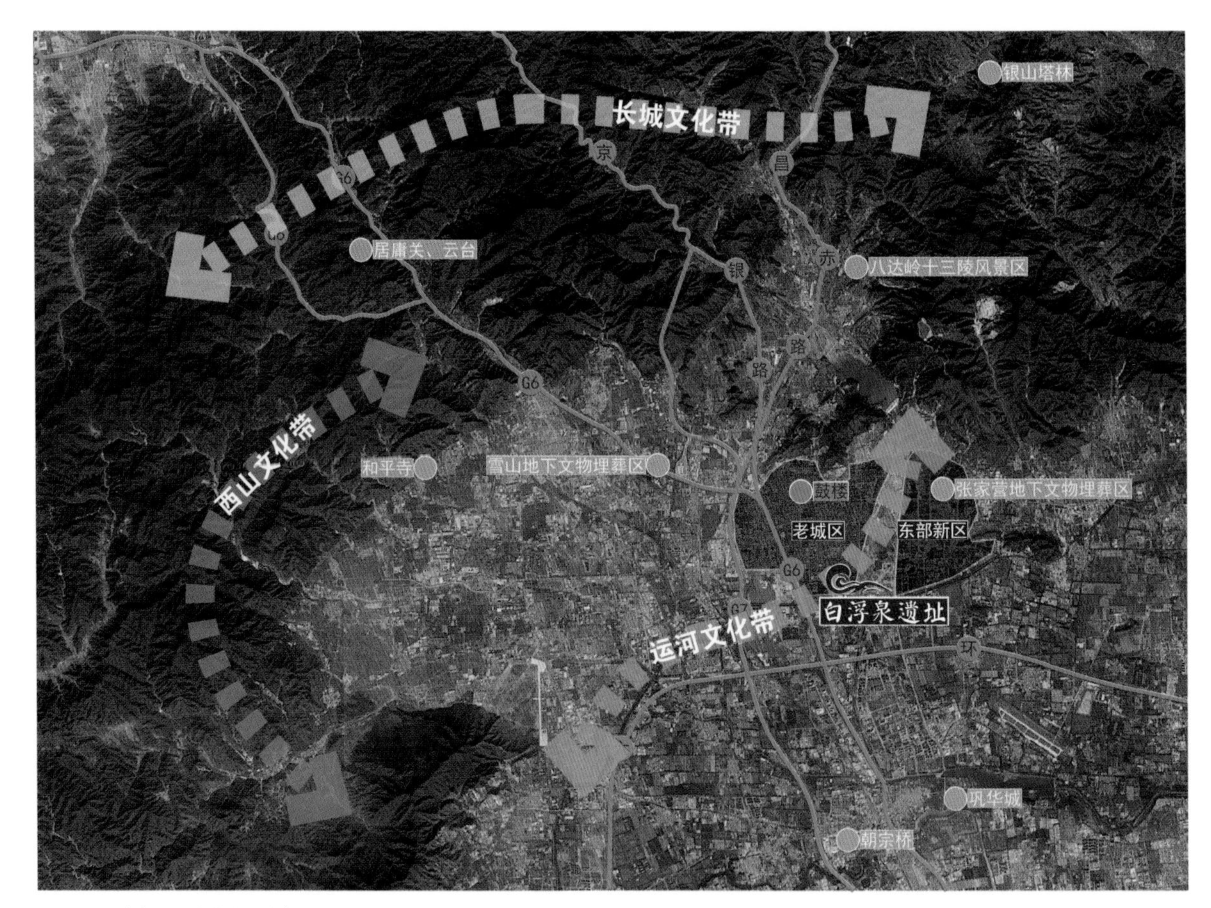

图 5-2 白浮泉周边地域文化分析

生息的遗迹。而据明确的历史记载，在距今2000多年的西汉时期，这里曾设昌平县治，隶属于上谷郡。昌平区境内有驰名中外的明十三陵、"天下第一雄关"——居庸关、十三级浮屠的辽代银山塔林等。全区现有文物保护单位78处，其中国家级重点文物保护单位4处，市级重点文物保护单位5处。昌平区由长城文化带、西山文化带和运河文化带环绕，这三个文化带相互依存，经过两千年的历史积淀，形成了完整的昌平文化内涵。

（二）昌平中心城区古城永安城的历史演变

城与关山共云天，悠悠相望五百年。山是军都山，关是居庸关，城即永安城，永安城是拱卫之城，是京师北部整体防线的战略节点。

1.永安城建成史与主要名胜古迹

明代以前，昌平县城位于今昌平旧县村。明代"土木之变"后，为了加强对皇陵的守备，在旧县城东八里处（1里=500米）建永安城，至今已有近600年历史。《光绪昌平州志》云："州故永安城。明永泰元年筑。东、西、南三门，俱重门。瓮城内、外各有层楼，周一千四百九十二九尺八寸，高二丈一尺。”永安城督造官，一种说法是由昌平侯杨洪担任，病逝后杨洪之子杨杰袭任昌平侯，继续建造永安城。

永安城初建为正方形，嘉靖十六年（1537年）在城南复筑一城与之相连，形成了南北长、东西窄的长方形。1956年城墙被拆毁。现在的东南西北四条环路以内的地域即为当年永安城的范围，今天的昌平中心城区就建立在永安古城的历史根基之上。

永安城的名胜古迹，首推谯楼（后称“鼓楼”）。“天顺三年，天寿山守备廖庸奏建谯楼于城中。”据历史记载，谯楼于天顺二年（1458年）动工兴建，于天顺三年（1459年）竣工。谯楼高约六丈，分为三层。

永安城内值得一提的还有永安四景，特别是“三山不露”。“三山不露”是八卦亭、文庙、县衙三处天然小山的合称，因为这三处小山的高度均不高又都在高墙深院之内，“三山不露”即因此得名。

2. 昌平城区的发展特征消失

在近代之前的中国城市建设通过建设城墙这个要素，将城市和周围的自然环境隔离、区别并作为异质的存在独立于自然之外。1956年城墙的拆除，使昌平原有的城墙环绕的礼制关系明确的城市形态不复存在，1971年具有512年历史的谯楼拆除，造成了昌平城市标志、文化支柱消失。快速城镇化和科技园区建设造成了历史积淀不足，新的城市特色未能形成。

三、白浮泉 · 化庄 · 龙山 · 凤山区域文化溯源

“白浮泉—化庄片区”位于昌平区南部，属昌平中心城区城南街道办事处。全面、深入地解析“白浮泉—凤山”片区的历史渊源、文化内涵，是本设计方案的首要任务，也是文化侧重的城市设计的重要设计手段。

昌平历史上被称作昌平州，在明朝的隆庆版、清朝的康熙版和光绪版三次州志中都对龙山、白浮泉有所记述，其中在清朝光绪版州志中以《昌平山水故城

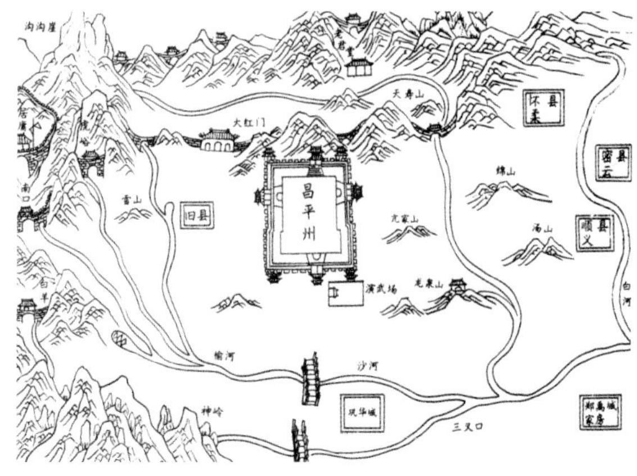

图 5-3 昌平州地图（1673 年）

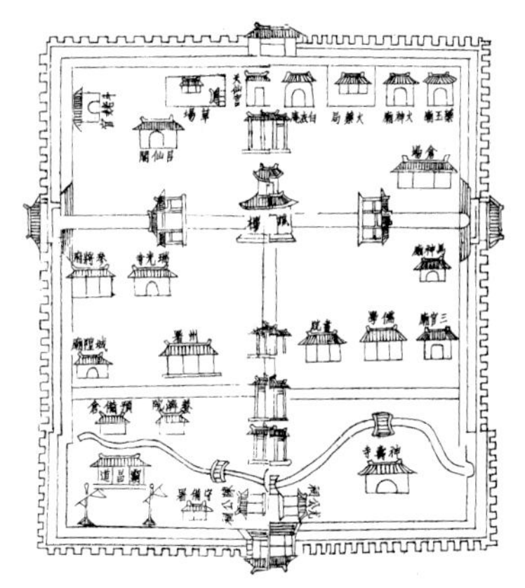
图 5-4 昌平州永安城地图

图 5-5 永安城谯楼，于 1971 年拆除

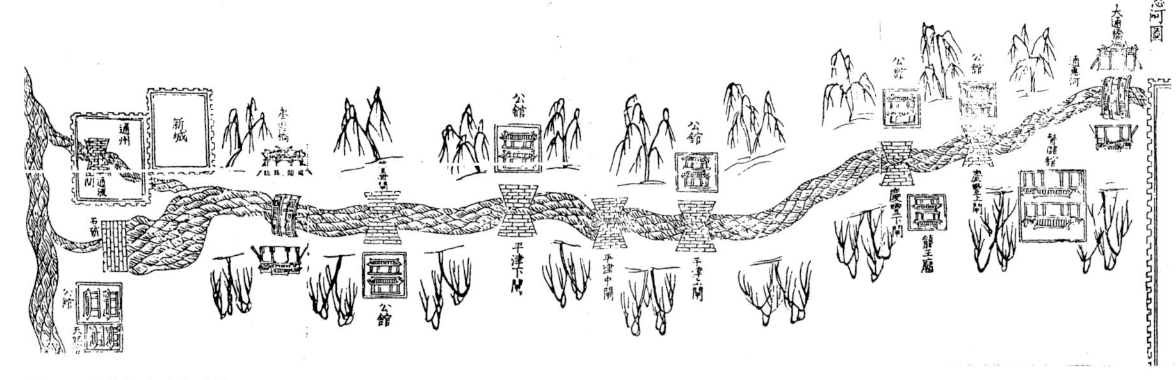

图 5-6 通惠河流域分布图

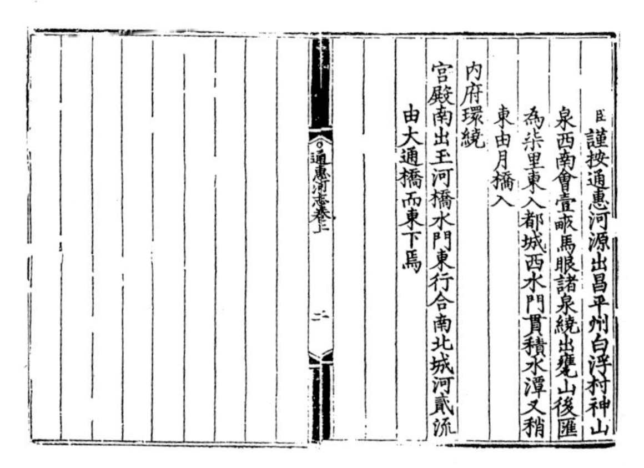

臣謹按通惠河源出昌平州白浮村神山
泉西南會壹畝馬眼諸泉繞出甕山後匯
為柒里濼東入都城西水門貫積水潭又稍
東由月橋入
內府環繞
宮殿南出玉河橋水門東行合南北城河貳派
由大通橋而東下馬

通惠河志卷上 二

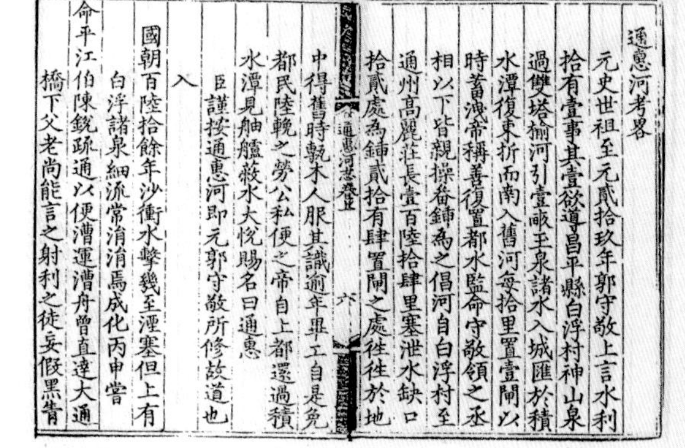

通惠河考畧
元史世祖至元貳拾玖年郭守敬上言水利
拾有壹事其壹欲導昌平縣白浮村神山泉
過雙塔榆河引壹畝玉泉諸水入城匯於積
水潭復東折而南入舊河每拾里置壹閘以
時蓄洩帝稱善復置都水監命守敬領之丞
相以下皆親操畚鍤為之倡河自白浮村至
通州高麗莊長壹百陸拾肆里塞泄水缺口
拾貳處為錘貳拾有肆置閘之處往往於地

通惠河志卷上 六

中得舊時甎木人服其識逾年畢工自是免
都民陸輓之勞公私便之帝自上都還過積
水潭見舳艫蔽水大悅賜名曰通惠
臣謹按通惠河即元郭守敬所修故道也
入
國朝百陸拾餘年沙衝水擊幾至湮塞但上有
白浮諸泉細流常涓涓焉成化丙申嘗
命平江伯陳銳疏通以便漕運漕舟曾直達大通
橋下父老尚能言之射利之徒妄假黑眚

图 5-7 通惠河考略

图》和《燕平八景图》之《龙泉漱玉》中以图画的形式予以记载。元代的《元史——河渠志》、明代的《长安客话》、明末清初顾炎武所写的《昌平山水记》、清代的《日下旧闻考》等历史古籍中都有对龙山、白浮泉的描述。“白浮泉即龙泉山也。”

（一）“运河北望、白浮为首”——大运河“源”点溯源

2005年12月15日，大运河申遗发起人（运河“三老”）：郑孝燮（古建专家，91岁）、罗哲文（文物专家，82岁）、朱炳仁（中国工艺美术大师，62岁）三老联名向运河沿线18座城市市长发《关于加快京杭大运河遗产保护和“申遗”工作的信》，引起大运河沿线城市的积极回应，就此拉开了大运河申遗的序幕。作为和长城一样的“线性遗产”，大运河申遗不是“全线申遗”，而是“点段申报”，涉及的河道总长只有1011千米。“点段申报”无疑会造成大运河文化挖掘不够全面，当然也给学界对于大运河文化后申遗时代的挖掘留下空间。

北京入选“两段”“三个遗产点”，“两段”分别为通惠河通州段、通惠河北京旧城段（玉河故道），“三个遗产点”分别是什刹海、澄清中闸、澄清上

闸。北京段作为京杭大运河北端的端点，有大量的大运河文化遗产，入选的“两段三点”显然对于大运河北京段的运河遗产尚有诸多疏漏，大运河北京段尚有诸多可挖掘之处。

根据申遗报告和《大运河遗产保护规划（北京段）》，大运河遗产北京段保护以元至明清的京杭大运河作为北京段大运河遗产保护的核心，并选择最具代表性的元代白浮泉引水沿线、通惠河、坝河和白河（今北运河）一线河道作为北京段大运河遗产保护的主线。

现存于龙山脚下的白浮泉遗址——九龙池是全国重点文物保护单位。九龙池当地人俗称九龙口，为当年白浮泉水所在地，现在经过修缮上面建了一座元代风格的亭子，上悬匾额“白浮之泉”。亭内有一石碑，石碑内容为侯仁之所写的《白浮泉遗址整修记》，其中说：“昌平城东南郊有龙泉山，或曰龙山。山麓有裂隙泉，昔日出水甚旺，即《元史·河渠志》之白浮泉，亦称神山泉。昌平沿山一带多有流泉，其为利之溥与历史上之北京城息息相关者，首推白浮泉。”侯仁之同时强调：“白浮泉是北京生命之根。”

元代熊梦祥《析津志》则称，昌平白浮泉为“京闸坝之源”，白浮泉是“元代京杭运河最北段的起源”。

可见白浮泉是“北京生命之根”“运河之源”。白浮泉作为“北京生命之根”和大运河“源”点的史实，不仅应当得到业界的承认，并应向公众普及，将其纳入“人文北京”建设的范畴。对大运河“源”的挖掘对于我国的后申遗时代运河文化完善具有现实意义。

图 5-8 从山脚下望都龙王庙

图 5-9 都龙王庙侧影

图 5-10 都龙王庙、白浮泉遗址及京密引水渠现状

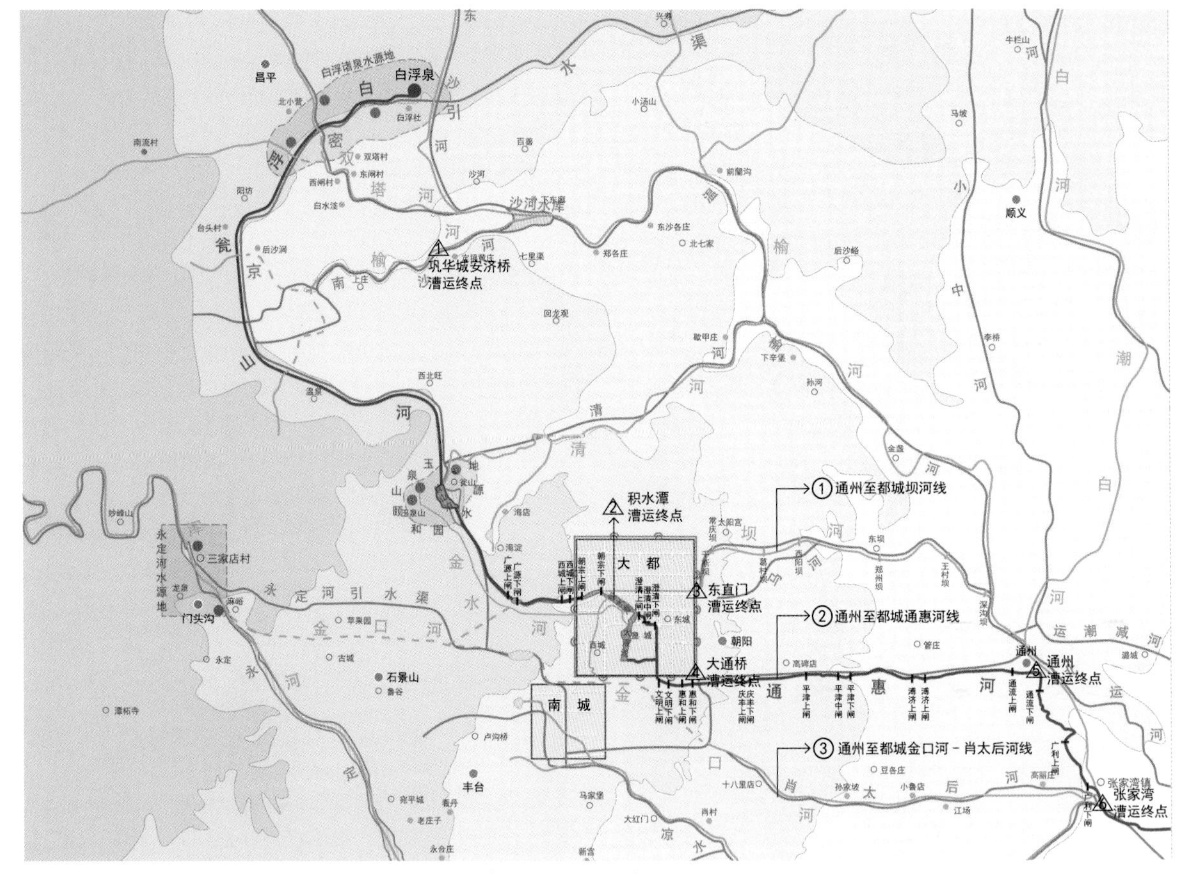

图 5-11 京杭大运河水源点及元通惠河（底图引自侯仁之《北京历史地图集》）

（二）白浮瓮山河与其他渠道

1977年，昌平文物队对于白浮泉到横桥村一段进行过考古挖掘，基本确定自白浮泉起的第一段河道就在今京密引水渠以北的200~300米处。同时将众多泉水引入白浮瓮山河的《元史》中"凿六渠，灌昌平诸水"，是沿途白浮瓮山河"支流"。2012年11日，在昌平马池口镇旮旯屯发现一处古水渠遗址，长约60米，宽约2米，最深处约2.9米，两壁和底部砌着大小不等的条石。根据考古意见，其很有可能是"六渠"之一。

最值得注意的一点是，在郭守敬的引水计划中他充分掌握了北京湾西部微小地形的变化，线路不是把白浮泉（海拔约54米）直接引向东南，而是向西至西山麓：然后大体沿50米等高线南下，避开河谷低地，沿途拦截沙河、清河上源及西山诸泉，再向东南注入瓮山泊。这个线路，以现在的技术水平来看，仍是最佳的选择。而现代水利工程京密引水渠，自白浮村至昆明湖这一段的线路，大体走向也基本与元代故道一致，仅小有调整，足证当初地形勘测之精确 。

图 5-12《龙泉漱玉》

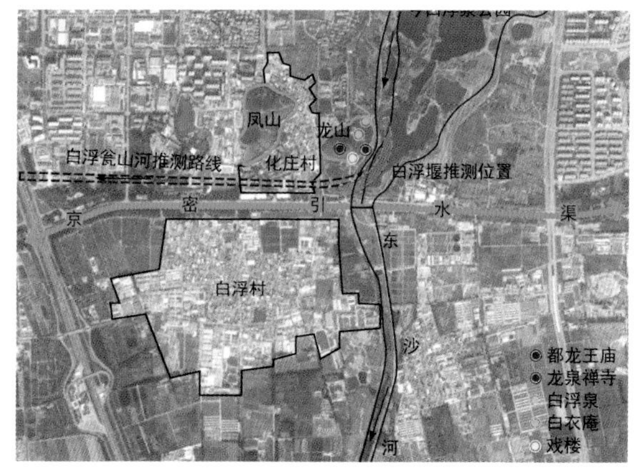

图 5-13 白浮泉龙凤山变化要素分析

（三）“龙泉漱玉”——都龙王庙、龙泉寺、九龙池等

由元朝郭守敬开凿京城漕运，确定白浮泉为北部诸泉之首开始，在历史的发展过程中于龙山上又陆续建设形成了九龙池、都龙王庙、龙泉禅寺、白衣庵、戏楼等。明清两代，白浮泉虽不再作为京杭大运河的主要水源，但仍然是昌平的重要水源和祈雨之地，承载了重要的社会功能和意义。

白浮泉遗址现存龙山、凤山两座山丘，龙山顶存都龙王庙，白浮泉旧址在1989年被重新修复。《光绪昌平州志》载：“城东南五里曰龙泉山，旧名白浮山。旧志：龙泉山即白浮山，在城东南五里。上有都龙王祠，山半一洞，有人梯石而下……”

在都龙王庙庙内于乾隆十七年（1752年）刻的《都龙王庙置田碑记》中记载：“……中有庙，翼然者三，一白衣庵，一龙泉寺，其峰顶则有都龙王庙焉。”龙王庙以“都”言之，谓其为京师第一龙王庙，盖因其自元代始为京杭大运河漕运之源。

《光绪昌平州志》载：“都龙王庙，在龙王山巅，明洪武八年重修，光绪四年祈雨有灵，奏请御赐‘祥徵时若’匾额重修殿宇。”匾额悬挂于大殿，供奉龙王神像，殿前立有多块石碑，为明清及现代的修缮碑；两侧有配殿和钟鼓楼。龙泉寺原址复建，现为龙山度假村之用。当年尚有白衣庵一座，遗迹不可考。根据《龙泉漱玉》图推测，位置应在龙山东北坡，九龙池上方。

龙山非物质文化遗产活动——“六月十三龙山庙会”，从农历六月十一至十三日举办三天。龙泉寺西南的平地上原来有座戏楼，现已不存，根据现场了解，应在龙山南麓。

（四）化庄、白浮村历史与龙山、凤山自然特征

史上关于化庄、白浮村的记载如下：“化家庄距城五里……西至凤凰山二里。”“白浮村距城八里……东北至龙王山二里……”有《晚留白浮山庄》诗记：“贫家偏好客，移席就山庄……遥见龙泉寺，龙灯水一方。”

“白浮山上有二龙潭，其水流经白浮村。元郭守敬筑堰……以便天下漕运。”可以说明泉水本来是流过白浮村的，元代郭守敬筑“白浮堰”截流向西。当时在白浮泉下修筑了堤堰，既可形成调节水池，又可避免东沙河水冲刷。

龙山（海拔约90米）、凤山（海拔约120米）不仅承载了厚重的历史文化记忆，同时也是昌平区主城区的自然空间制高点，其扮演的城市空间节点角色也尤为重要，是城市空间发展的视觉焦点。

图 5-14 区域交通分析图

（五）"白浮泉、化庄、龙山、凤山"的文化总结

概括"白浮泉、化庄、龙山、凤山"区域的历史，我们首先看到：元朝第一次使北京成为整个帝国的都城，如何引水济漕使之能充分供应京城皇家用水与日常百姓用水一直是几百年来帝国首都最为关键的问题。白浮泉对于北京在水利学和北京建城时的意义在于，它巩固了北京第一次作为整个帝国首都的基础，自此北京才开始了几乎是没有间断的国都史。大运河"源"头文化无疑是该区域文化的核心要素。

泉首先提供了水，既保障了生活，也造就了泉景；泉因为灵应，先后促成了两座祠庙的建立，引起了各阶层的关注；祠庙因为"灵应"，又增强了泉的地位，并为泉增添了新景。都龙王庙、龙泉寺、白浮泉（龙泉）位于一座小山上，恐非偶然。"白浮泉—龙山"整体作为中国大运河文化遗产的一个重要节点， 从最初单纯的泉景、水利设施逐渐生发出相关的建筑和景观，成为了一处综合的文化景观，凝结了丰富的内涵和意义。

著名古建专家罗哲文曾说："如果没有这条运河，北京城就可能修不起来了。"大运河是"历史之河，文化之河"，从各种意义上估价它的水文化与城市发展的文化遗产内涵都有意义。著名学者侯仁之强

01 都龙王庙
02 龙泉禅寺
03 白衣庵
04 九龙池
05 兴昌阁
06 现状企业
07 传统文化居住区
08 回迁安置区
09 公交车站
10 小学
11 幼儿园
12 社区服务中心
13 派出所、治安支队
14 现代商务办公区
15 戏楼
16 白浮瓮山河遗址展示

图 5-15 城市设计总平面图

调说："与历史上之北京城息息相关者，首推白浮泉。"可见，运河北望，白浮为首。

四、构建文化昌平、山水昌平——文化侧重型"白浮泉、化庄、龙山、凤山"区域城市设计

"白浮泉—凤山片区"位于运河文化源头，与另外两个文化带唇齿相依，"白浮泉、化庄、龙山、凤山"区域城市设计以文化挖掘与规划为引导，以龙山、凤山、京密引水渠、东沙河公园、都龙王庙、白浮堰、白浮泉—九龙池、龙泉禅寺、白衣庵（娘娘庙）、戏台、白浮渠旧道等复兴昌平地区历史文化精华，以白浮源大运河文化、"龙山"为文化城市重塑主线，基于大山水研究构建亢山、龙山、凤山"三山入城"的新"山""城"关系，山、水、城、阁共同构建文化、山水昌平，是文化侧重的城市设计的实践。

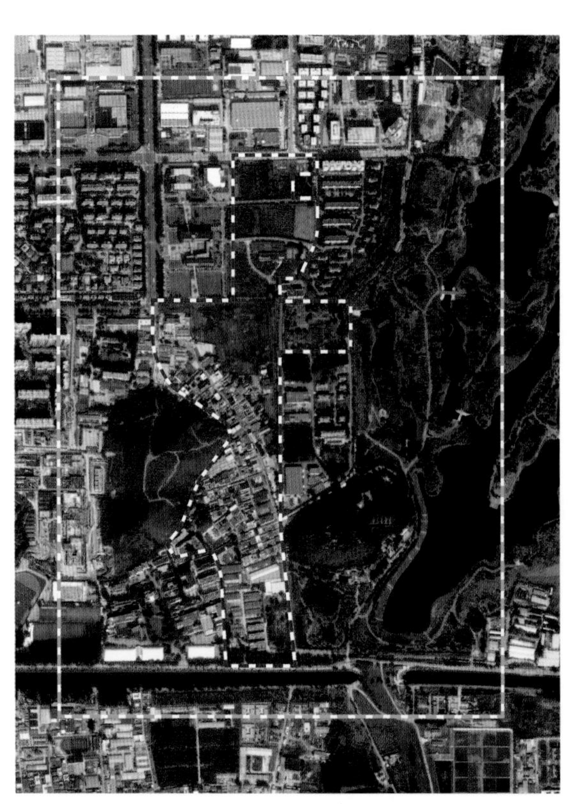

图 5-16 规划范围图

（一）构建昌平文化精华地区

建设文化精华地区的目的在于，选择具有突出的文化价值和历史价值的重点地段，按照历史联系和空间关系进行整体生成的综合集成，塑造有特征的历史人文景观和自然景观的历史文化区域，与历史文化名城的地位相称，是重构文化城市的重要途径之一。

以大运河之源文化为引领，通过"一泉二庙楼、两山两水两村"的保护性修葺、复建等，重现燕平八景之一的"龙泉漱玉"，实现构建昌平历史文化精华地区的总体构想。

"一泉"即龙泉漱玉：昔日，九龙口旁古槐参天，绿柳垂荫，清泉吐着银花，与青山古树相映成趣，被文人们列入燕平八景，雅称"龙泉漱玉"。此次公园建设，旨在对白浮泉进行整修，再现这一景观。

"三庙"即都龙王庙、龙泉禅寺、白衣庵：对都龙王庙和龙泉禅寺进行保护修缮，重建白衣庵。白衣庵和戏楼在昌平历史州志及相关典籍中均有记载，作为龙山历史文化体系中重要的组成部分，两者的缺失却为一种遗憾。规划提出复建白衣庵和戏楼，复建需通过严谨细致地探究两者的所处位置、建筑形制特点及现状建设条件进行设计实施，最终完善龙山、白浮泉的历史文化体系节点。

"一楼"即古戏楼：在都龙王庙山门下恢复古戏楼，建设"水"文化主题广场。

"两山"指龙山与凤山：将龙山与凤山通过水系、步道、绿化带有机连接，遥相呼应。

"两水"指东沙河与白浮瓮山河（现京密引水渠）：将东沙河（现昌平滨河森林公园）与白浮瓮山河（现京密引水渠）的景观进行整体规划，形成京北地区一条富有大运河文化底蕴的园林景观绿化带。

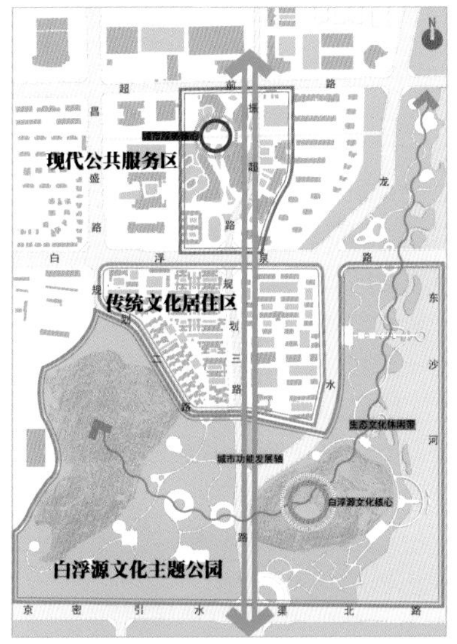

图 5-17 空间结构规划图

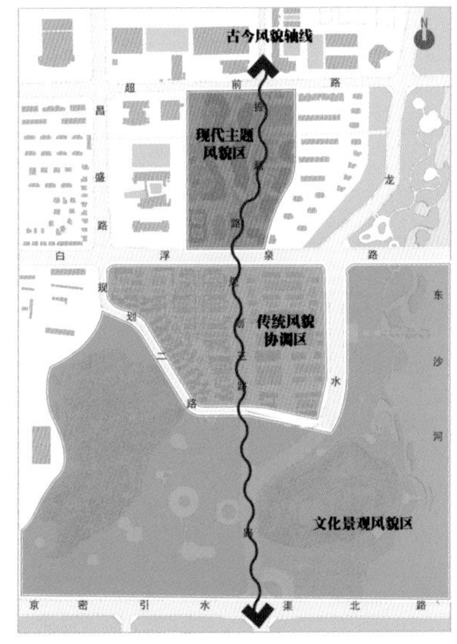

图 5-18 街区风貌分区图

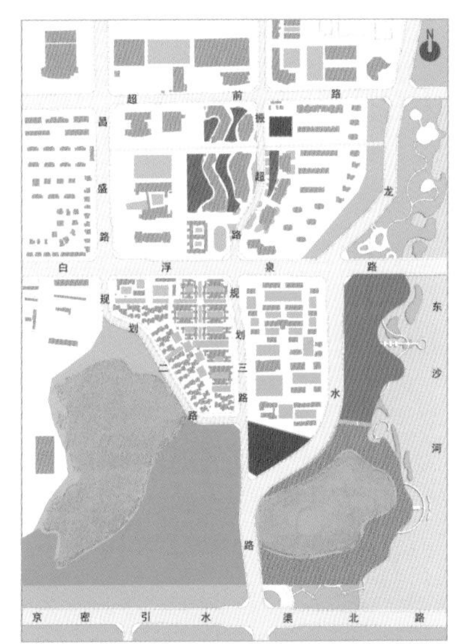

图 5-19 开放空间规划图

“两村”指化庄村与白浮泉村：对化庄村进行棚户区改造，对白浮泉村进行有机更新，强化龙山周边村落主要是化庄村、白浮泉村与大运河—白浮泉遗址的自然关联。

（二）建设大运河源头——白浮“源”遗址公园

在积极践行保护、传承、利用白浮泉龙凤山的总体规划思想下，为北京生命之根做出新的历史探源的基础上，展现昌平文化核心内涵，再塑历史上白浮泉的辉煌。借此规划，提出打造以白浮泉为文化主题核心的白浮源文化休闲生态公园，形成“一源”回顾历史文化、“一园”惬意休闲生活的白浮图景。

梳理白浮泉乃至北京漕运对于大运河的重要地位，考证重要节点及古水道走向，根据考古挖掘，保护白浮泉—瓮山河遗址，重点保护现存的白浮泉—九龙池遗址以及都龙王庙，制定政策、修缮、管理三位一体的保护措施，并建议部分占据文物的单位迁出，最终达到真正意义上的历史文化保护目的。

（三）筑兴昌阁——昌平文化新地标

《城市意向》中，凯文林奇曾描述：“一个不论远近、速度高低、白天夜晚都清晰可见的标志，就是人们感觉城市复杂多变时所依靠的稳定的支柱。”凤山并无历史文化遗存，但它比龙山高30米，是整个昌平主城区的空间制高点。鼓楼拆除多年，如何恢复城市视觉地标和心理支柱也是设计目标之一，故规划提出于凤山山顶制高点兴建“兴昌阁”。兴昌阁的建设，正如雷峰塔之与杭州、黄鹤楼之于武汉、玉泉塔之于颐和园，期冀能形成昌平新的文化地标，代表着昌平的未来。

（四）山、水、城、阁——构建“山水昌平”

昌平城墙、鼓楼的拆毁已不可恢复，只为遗憾。城墙的拆除，打破了原有城市的格局和发展界限，改变了昌平城市的发展轮廓，打破了城市和城外的龙凤山景区的界限，城市的发展重新构建了和原城外的亢山、龙凤山的关系，先是亢山进入城市，随着昌平科技园的建设，龙山、凤山、白浮泉规划的实施，龙山、凤山将与亢山共同将进入城市，与城市融为一体，规划将试图重新构建"新昌平三山"，从永安城的"三山不露"再到"三山入城"，与东沙河公园、京密引水渠、兴昌阁共同重构新昌平城区的山水文化和山城关系——即"三山、两水、一阁、两城"的山水昌平。

（五）龙山文化、白浮泉文化等复兴工程

白浮泉的历史文化传承离不开非物质文化的传播，项目策划重启久负盛名的农历六月十三龙王庙庙会，复建戏台，复兴传统演出，展现种类繁多的非物质文化遗产的传承，形成以白浮泉文化休闲生态公园为载体，历史记忆与现代休闲相结合的节事活动体系。

建议以白浮泉为起点，沿京密引水渠路线，承办"半程马拉松"或"全程马拉松"比赛。增强白浮泉的宣传，弘扬大运河、龙山、白浮泉文化。

（六）城市设计实施要点

依据龙山白浮泉的文物保护规划及区域控制性详细规划，严格管控龙凤山周边的城市建设。本项目规划控制分为历史文化控制区、控制协调区及外围城市建设区。历史文化控制区以龙凤山山顶最高点为原点，200米为半径，控制协调区为半径200米至半径600米，200米至300米半径范围内建筑限高6米，300米至400米半径范围内建筑限高12米，400米至500米建筑限高18米，500米至600米半径范围内建筑限高24米，形成阶梯递进式的高度空间控制。

图 5-20 南向北鸟瞰图

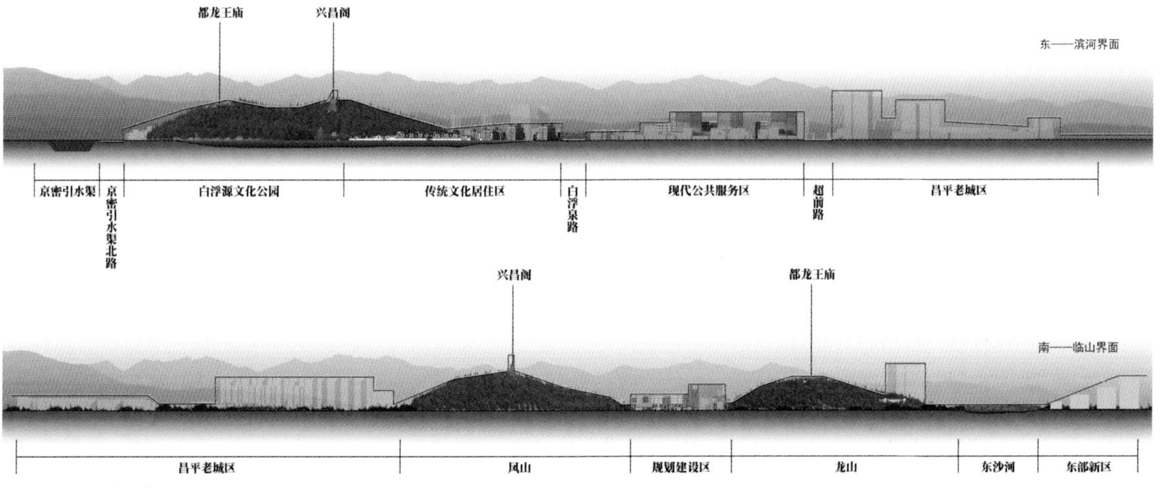

图 5-21 天际线分析图

图 5-22 西北向东南透视图

图 5-23 从京密引水渠看龙凤山效果图

图 5-24 白浮泉地区航拍实景

东沙河

图 5-25 规划鸟瞰图

低密度住宅
现代服务区

图 5-26 建议控制范围示意图

本项目打造南北向城市空间轴线、历史时间轴线、城市公共服务功能轴线。规划形成由南至北、由历史风貌至现代风貌的过渡、融合、交汇的城市空间，形成规划区南部的白浮源文化公园、中段的传统文化生活区、北部的现代公共服务区。以龙山、凤山为基础，打造生态自然、环境优美、舒适惬意的昌平主城区的南大门，最终建设成为独具文化特色的城市形象识别区域。

五、结论与展望

城市是人类积累性创作的结果，城市风貌的形成源于其发展过程中的积累性遗存，快速发展造成的破坏广泛存在且经常不可避免，城市随着历史的发展都经历着不断的"建构—破坏—重构"的过程。从城市化快速发展的现实来看，昌平中心城区经历的在中华

图 5-27 西南向东北规划鸟瞰图

图 5-28 城市景观效果图

图 5-29 城市景观效果图

图 5-30 城市景观效果图

人民共和国成立初期的拆除城墙运动和改革开放后快速发展后造成的景观破坏、特色消失、文化湮灭在我国广泛存在。

近年来，城市文化建设和重构的重要性越来越为人们所重视。加强探索和发展富有特色的城市文化，是推动从功能城市走向文化城市的一条重要而可行的路径 。2013年中央城镇工作会议提出："依托现有山水脉络等独特风光，让城市融入大自然。"

"北京昌平白浮泉·化庄·龙山·凤山区域城市设计"方案依托对悠久历史记忆和白浮泉大运河"源"文化的的挖掘，通过建设以白浮泉为核心的历史文化精华区域，"三山、两水与两城"共融重构山水城市，通过激发"活态遗产"认知并展开相应历史遗产文化活动，使文化遗产湮灭特征消失的昌平中心城区通过文化复兴的引领，重建文化城市、山水城市的目标得以初步实现。

项目信息及版权所有

"白浮泉、化庄、龙山、凤山"城市设计
业主：北京兴昌高科技发展有限公司
建设地点：北京市昌平区
规划设计：北京建院约翰马丁国际建筑设计有限公司
设计团队：1.朱颖 2.张达志 3.刘玉锋 4.乔崎 5.白雪 6.邢屹 7.梁学强 8.林新业 9.《中国建筑文化遗产》编辑部
规划面积：180公顷
设计时间：2017年
摄影/图片版权：赵欣然 殷力欣 等/北京建院约翰马丁国际建筑设计有限公司
撰文：朱颖

陆记 Note Six

历史昭鉴未来
History Mirrors the Future

哈尔滨731士官宿舍区遗址保护
规划竞赛方案

Competition Scheme for the Protection Plan of
Harbin 731 Site

本记以城市设计的灾难历史观展示城市设计的空间、风景线与环境特性，
既是对建筑师职责的再认识，
也是弥补建筑师跨界文博“空白”的一个尝试。
城市设计作为提升城市建设管理水平的重要工作，
在城市更新和对城市空间修补上的作用独特，
从而与人们熟谙的生命、宜居、平安反思等命题同构。

This paper tries to integrate the presentation of disaster history with the city's space planning, landscape and environmental attributes. It offers an opportunity to rethink the obligations of architects and also an attempt to fill the gap in architects' involvement in cultural museum. As an important tool to improve the city administration level, urban design takes a unique part in city's renewal and spatial fix. In this sense, it goes in parallel with many well-known propositions concerning life quality, livability and reflection on wellbeing.

哈尔滨731士官宿舍区是日本帝国主义侵华行径的重要历史罪证之一，项目规划正值反法西斯战争胜利70周年之际，随着奥斯维辛集中营的“申遗”成功，“二战”时期的历史罪证遗迹在不断被重视，其历史价值、罪证价值及和平的教育意义极其重要。同时遗址区位于城市核心的平房区，经过数十年的城市发展，该区域所呈现的现状在城市形象、城市空间、城市交通及城市功能等方面都制约着城市的良性可持续发展。如何处理好历史保护与城市发展更新的关系，让历史遗迹活态传承发展是项目的核心问题。

项目基地位于哈尔滨市平房区侵华日军731部队罪证遗址历史文化风貌区西南角，东临公社大街、南临友协西三道街、西临新银街、北临新疆大街，规划用地面积5.7公顷，属于侵华日军731士官宿舍旧址的一部分（士官宿舍旧址东临友协大街、南临建文街、西临新银街、北临新疆大街，规划用地面积19.8公顷）。依据《哈尔滨历史文化名城保护规划》和《侵华日军第731部队旧址保护规划》，本项目属于建设控制地带，但不属于731 遗址核心保护区。

一、“731”的历史

（一）731部队概况

1933年8月，日本大本营在哈尔滨市南岗宣化街和文庙街的中间地带秘密设立细菌战基地，即“石井部队”，对外称“关东军防疫给水部”。1936年春季，石井部队在平房设营驻扎。1938年6月，日本关东军将石井部队营区、日本空军8372部队营区和平房镇划为特别军事区；同年9月，“关东军防疫给水部”大部分人员、设备迁至平房本部，这里成为世界上规模最大、设施最完备的细菌战基地。1941年，本部番号定为满洲第731部队。1939—1942年，日军在其侵占的北平、南京、广州、新加坡等地建立细菌部队，731部队成为日本细菌武器研究中心和细菌战策划指挥大本营。1939—1945年，至少有3000多名中外人士通过“特别移送”被转移到731部队当作“实验材料”，惨死于各种细菌试验和活体试验中。1945年8月10日，731部队接到提前撤退命令后，开始全力销毁资料，砸毁仪器设备，炸毁建筑，并用毒气杀害“特设监狱”中的在押人员后焚尸灭迹。8月14日，731部队组织人员乘火车南逃，各支队也自行撤退。8月15日，日本宣布无条件投降。8月18日，731

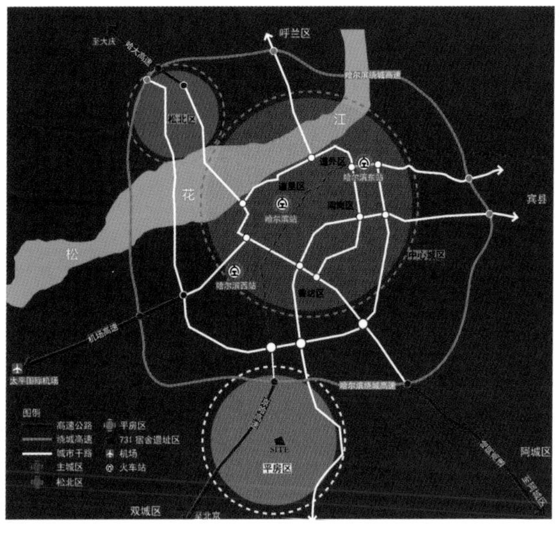

图 6-1 平房区区位图

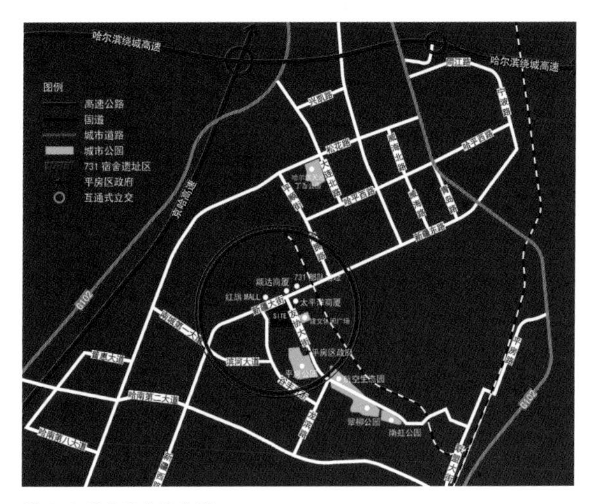

图 6-2 项目用地区位图

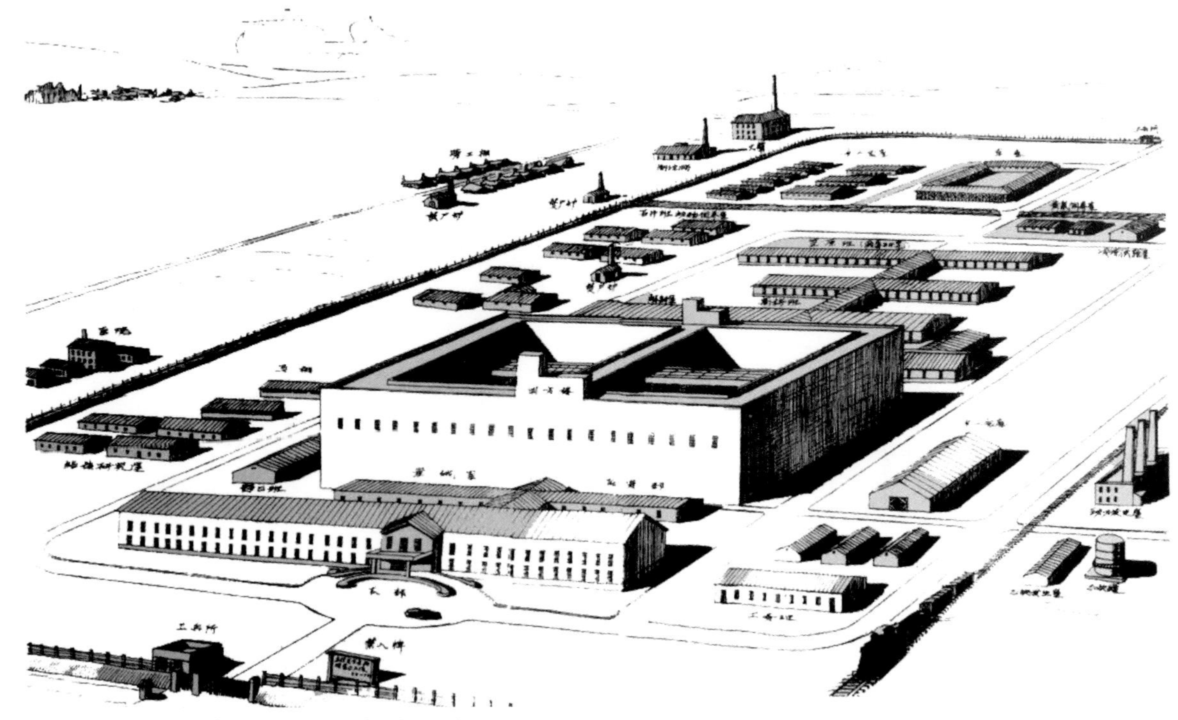

图 6-3 731 部队本部平房营区史记风貌图（罗克敏 绘）

部队绝大部分人员逃回日本。

（二）731部队旧址

1938年6月30日，日本关东军司令部发布命令，将731部队本部平房营区周围约40个村屯120平方千米的地域划为“特别军事区”。其中平房营区周边6.1平方千米的“甲号地区”为“无人区”，其余区域为“乙号地区”。731部队平房营区有建、构筑物80余处，功能分区明确。营区核心区域建有“日”字形土墙，东西长约614米，南北长约1000米，墙高2.5米，墙上架设高压电线，是“特别军事区”的第一道陆上防线。土墙内是以本部、细菌试验室及特设监狱为中心的731部队细菌研究、实验、生产的核心区域，用地面积约61公顷。核心区域以东设有航空指挥所、气象台、机场等。核心区域以南为教育部，是731部队培养细菌战后备力量的少年队的驻地。核心区域西南为731部队生活区（宿舍区），配有包括宿舍、神社、广场、运动场、花园、医院、学校等在内的完备生活设施。“甲号地区”“乙号地区”外缘分别设置第二、三道陆上防线，另设空中防卫线一道，包括日军其他部队军机在内的任何飞机不经允许禁止飞越其上空。日军还铺设了铁路专用线，为731部队输送物资、人员和“特别移送”的被实验者，并把731部队生产的各种细菌、培殖的母本动物和专用设备发往各地。除平房营区外，距营区西南4千米的城子沟野外实验场和位于南岗的细菌弹壳制造厂也是731部队生产、实验细菌武器的场所。教化街营外宿舍是731部队在哈尔滨市区内的营外驻地。吉林街联络处曾是石井四郎在哈尔滨市的住处，该建筑和日本领事馆都曾作为731部队实施

"特别移送"的转运站。

（三）历史沿革

日据时期——1936年，石井部队进驻平方扎营，建立了臭名昭著的731细菌部队，同时建设731士官宿舍区。宿舍区是部队生活区，分为单身宿舍及家属宿舍，基本以进深6米、开间4米的模式建设，是那个年代的建筑实物罪证。

哈飞时期——中华人民共和国成立初期，哈尔滨作为振兴中国工业的老工业基地，发展了一批的国家级工业企业。哈飞（即122厂）将士官宿舍区利用为职工宿舍，在原始建筑空间的基础上，顺应居住功能空间需求，破坏了一些原始的建筑空间，使历史罪证有了初步的破坏。

房改时期——随着房改时代的到来，士官宿舍区基本以户为单位产权转为私有。每户家庭根据各自不同的居住需求对建筑内部及外部空间进行了第二次改扩建，其中主要以增加外廊，扩建室内使用空间为典型问题，进一步破坏了历史原有物证的原始性。

现在——近些年，经济迅猛发展，市民生活方式改变，城市不断更新，市场经济活跃。平房区商业中心位置的士官宿舍区也随着经济的发展而产生变化，士官宿舍建筑功能也在由原始的居住功能向复合功能业态转变，在这历史的当口我们有责任和义务保护那段历史遗存。

二、"731"士官宿舍区遗址区保护与利用的总体构想

我们对731部队士官宿舍遗址区的处理方针，表达了对历史遗迹的保护与利用的关系理念：

站在高处，让历史昭鉴未来，
历史记忆的老照片，
平房未来的新名片，
展示线路与生活路径的剥离与叠加，
固定罪证和重筑空间的剥离与叠加，
牢记历史和追求美好的剥离与叠加。

三、"731"士官宿舍区的保护与发展策略

（一）综合评判文物性状，分层次制定保护利用原则

原貌保护原则——731士官宿舍区是反法西斯战争的物证遗存，不仅具有珍贵的历史保护价值，同时也有向后人展示的重要意义。规划选取基地局部几栋建筑作为历史原貌的展示，回顾那段历史。建筑空间不植入任何功能业态，只具备单一的展示历史功能。

保护性利用原则——基于现状保护建筑的使用功能，为了满足城市产业功能需求，基本延续现状业态功能使用情况，在以保护原则为基本要求的基础上，植入可适用于现状建筑内部空间结构的业态功能，实现保护为主、利用为辅的策略引导。

利用性保护原则——根据实地踏勘现状保护建筑，对改建程度较为严重的保护建筑，在保证建筑主体结构的完好性的基础上，实行以功能利用为导向的保护原则，如根据业态功能对空间进行需求规划的微小型办公及酒店等。

（二）多维度空间视角叠加，实现不同情感交融与碰撞

为了更好地记忆731士官宿舍区历史，全面展示历史风貌，规划提出从不同空间维度，创意设计栈道

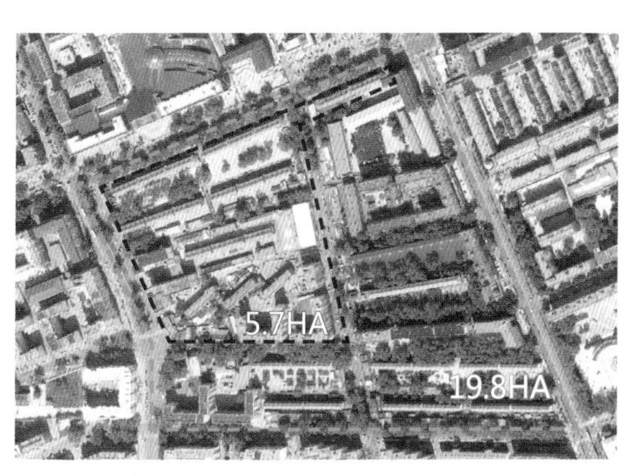

图 6-4 设计范围图

展示路径及高空观景平台，形成人视、半空间及高空多个空间视角，使观览者能够从建筑细部到整体规模形成全方位立体化的视觉感受。

（三）梳理整合既有业态，完善丰富城市产业功能

依托平房区战略产业政策，结合项目产业功能区域协同优势，梳理现状产业功能业态，迎合新时期产业发展新趋势，731士官宿舍区集合了商业服务、信息服务、科技创新服务、金融、展览展示、文化创意、居住休闲等功能，规划展示空间、沿街商业、地下商业、小型办公、文化创意、主题公寓、酒店等业态形式。

（四）明晰分解城市需求，创造多样差异化空间

步行街——规划分时步行街，打造吸引步行的休闲路径，形成项目的核心标志性空间之一。步行街是承载展示路径、休闲路径、商业路径、生活路径及沟通地上地下空间的复合性功能空间。

核心广场——打造项目的核心空间节点，是历史展示的核心集散空间，是规划展馆的景观形象空间，是地下商业的核心门户空间，是项目附近区域市民休闲的重要空间场所，是空间拓展的核心之一。

地下空间——基于项目基地的保护区属性，现状基地可利用空间相对有限。地下空间的开发和利用不仅规避了空间的制约性，也顺应哈尔滨城市寒冷的气候特征。设计规划地下商业空间及地下立体停车两

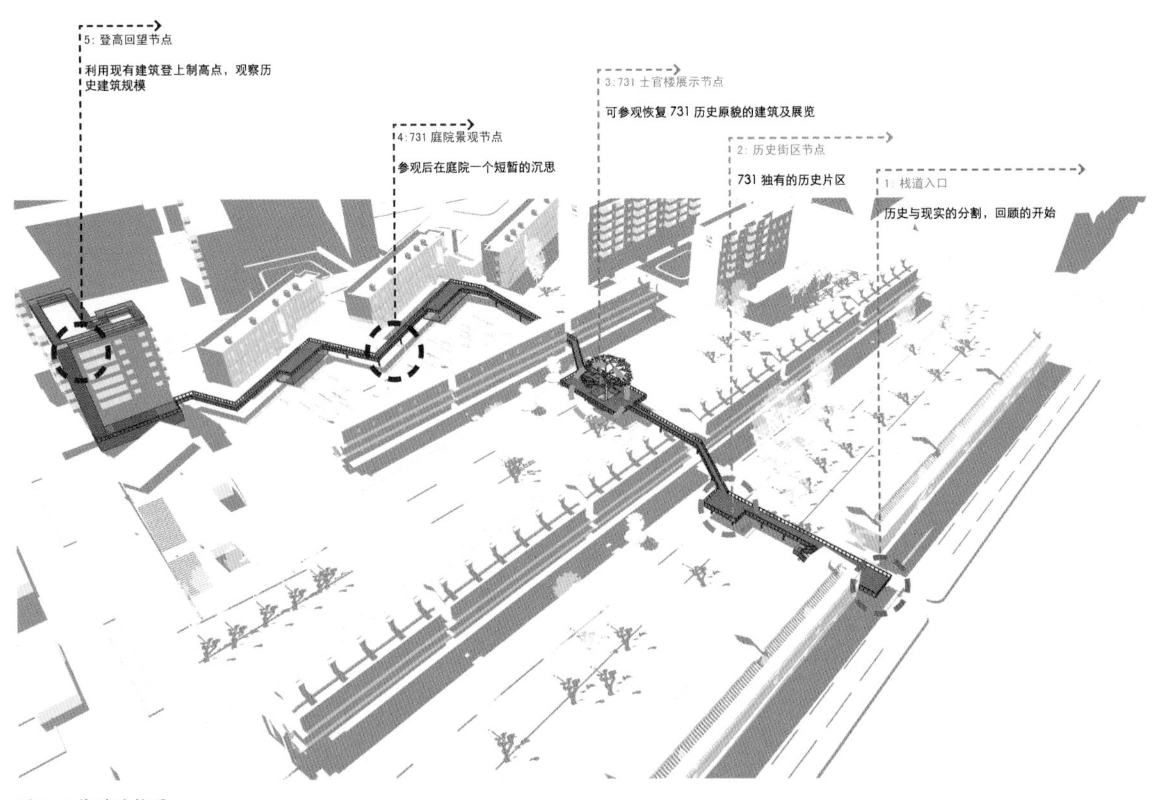

图 6-5 栈道结构图

图 6-6 栈道入口节点图

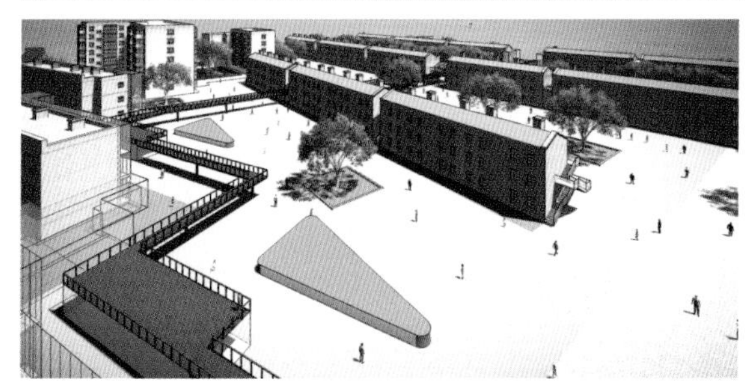
图 6-7 空间效果图

部分。地下商业空间不仅承接拆迁商业补偿的空间使用，也是项目作为城市核心商业区的地位体现。地下立体停车空间在合理配置静态停车需求的同时，对区域静态交通的规范化将起到良好的推动作用。

（五）全程设计运营机制，实现项目良性发展

保护为主、开发为辅——规划确定了“保护修复为主，开发改造为辅”的指导思想并成为基本运营开发的核心。

合理安置、保留权利——关注原住民的合理需求，保障其迁出后的生活、就业等诸多需求，对想重新回归的原住居民保障其回迁的权利。

业态遴选、设置门槛——为保障区域业态合理发展，招商引资时明确所需业态，设置符合区域发展的业态门槛，满足产业的良性发展。

产业循环、良性发展——严格控制区域的的开发建设。在引进项目产业时，要保证整个片区的可持续发展性，保持一定项目的持有性，保证收益的持续化。

四、“731”士官宿舍区的体系构建

（一）空间架构体系

“两轴”：规划延续基地的十字形的空间结构，打造南北纵向空间主轴及东西横向空间次要轴线。秉承项目规划理念特征打造纵向主题步行街，形成以游览、商业、休闲为主导功能的核心轴线。横轴是城市生活功能的次要轴线。“一心”：对现状5层居民楼及2层商场进行腾挪安置，规划以731历史、平房区历史以及平房区工业发展历程为主题的平房区展馆，形成区域核心，并拓展公共开放空间，有效地衔接融合历史展览空间及城市生活空间，形成项目的空间核心。“四节点”：规划形成731士官宿舍保护区北侧主要入口及西侧次要入口，南侧对接建文街，东侧对应建文休闲公园的四大空间节点。“多片区”：包含多种功能的项目空间片区。

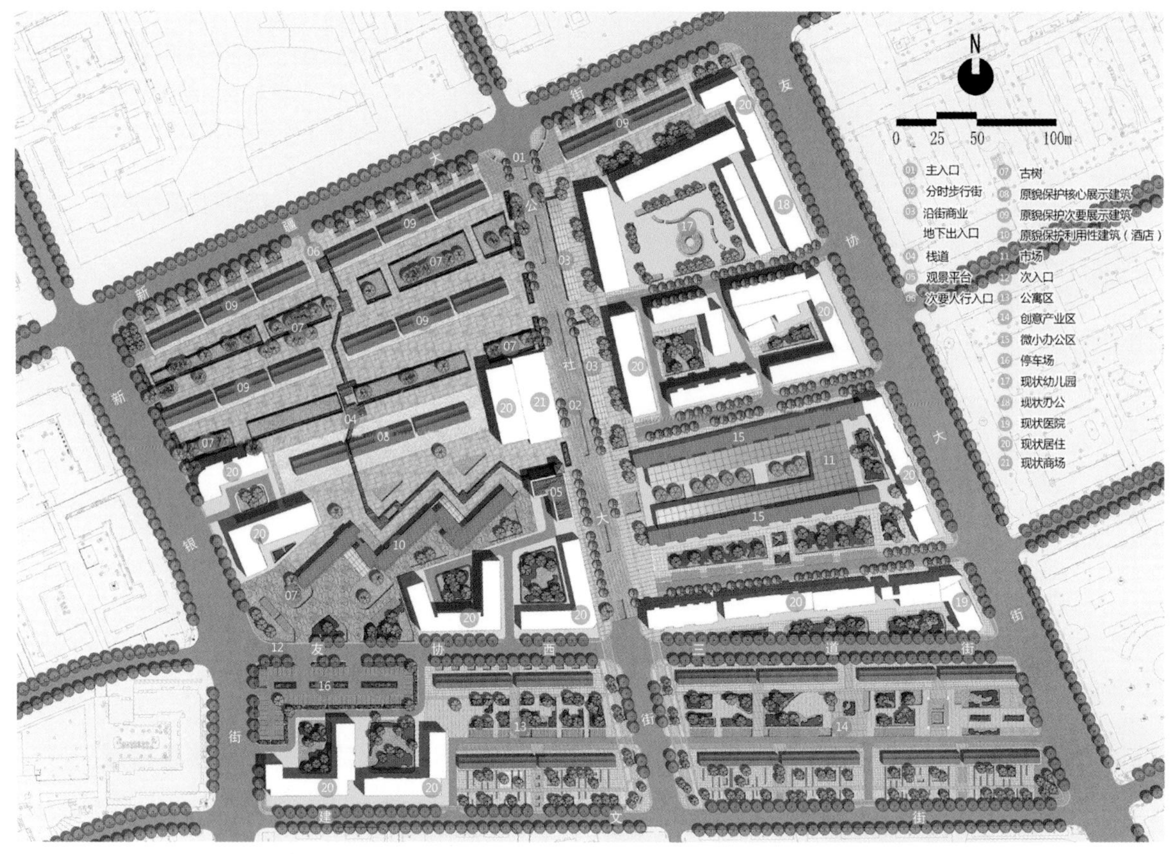

图 6-8 总平面图

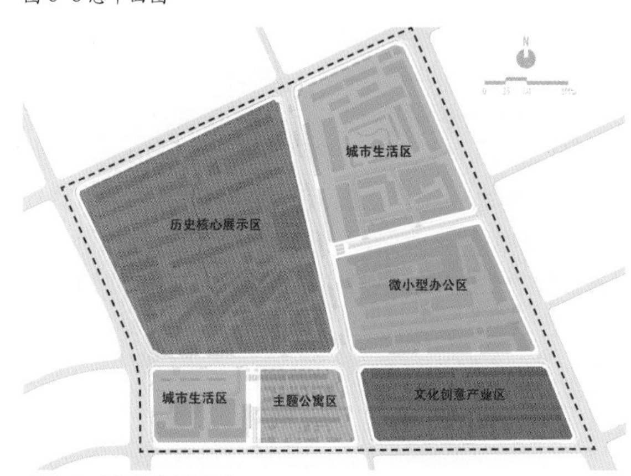

图 6-9 功能分区规划图

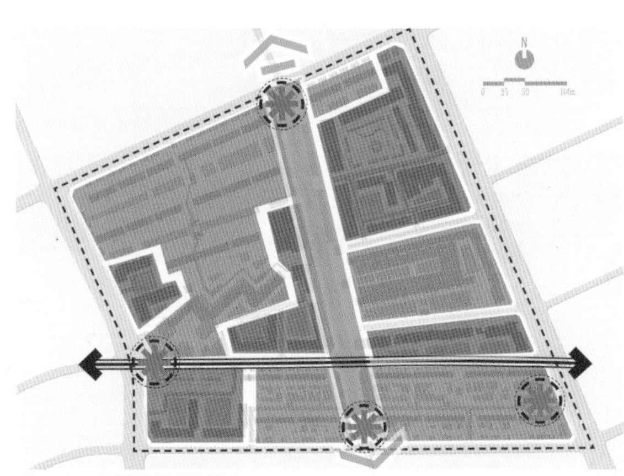

图 6-10 空间结构规划图

图 6-11 东北向西南整体鸟瞰图

规划兼顾项目的历史保护及城市功能的双重角色特征，形成历史核心展示区、微小型办公区、文化创意产业区、主题公寓区及城市生活区五大功能片区。历史核心展示区：即项目核心功能区，也是项目启动区，以731历史展示为核心功能，辅以商业、办公、公寓、酒店、休闲及城市生活等功能。微小型办公区：现状历史建筑被改造程度严重，规划以利用性保护历史建筑手法为原则，植入科技型、创新型、生态型小型办公功能，原址改造提升现状市场形象。文化创意产业区：依托平房区创意产业发展政策优势，以保护优先、利用为辅的原则，打造以艺术、创意、动漫、设计等相关行业为主题的创意产业区，形成平房区乃至哈尔滨市的文化聚集区。主题公寓区：利用现在历史建筑空间，以保护优先、利用为辅的原则，打造以“二产”产业配套的专家型公寓及城市生活配套的单身公寓等功能。城市生活区：为现状城市居民工作居住空间。

（二）历史展示体系

731士官宿舍区是日本侵略者的实物历史罪证，

图 6-12 731 展览馆效果图

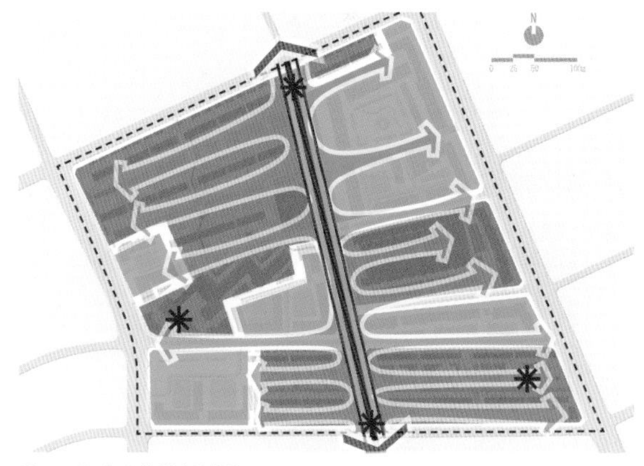

图 6-13 景观结构规划图

图 6-14 三联楼改造效果图

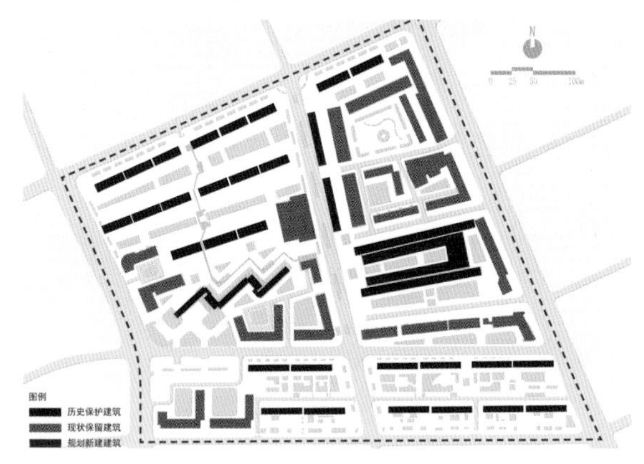

图 6-15 建筑保护规划图

图 6-16 沿新疆大街改造效果图

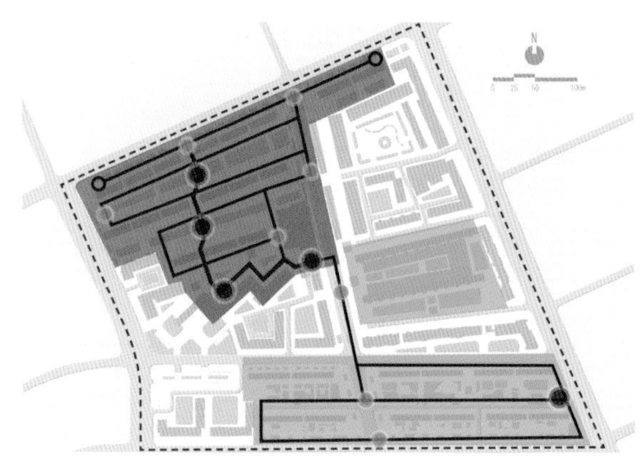

图 6-17 游览空间规划图

爱好和平的人每年来此络绎不绝，他们通过游览追忆那段罪恶的历史。此次规划调整将为世人展现一个多角度的历史实证。将总体宿舍区规划为核心游览区（启动区）与外围游览区两个空间区域；设计普通地面游线及特色栈道游线两种路径；打造地面人行视角、栈桥半空视角及楼顶高空视角三个维度的空间感受，全方位立体式地感受那段历史。

（三）建筑保护利用体系

根据文物保护的要求、基于对现状城市功能建筑的尊重及对未来城市功能建筑空间的需求，对基地内部建筑采取绝对保护、保护性利用、利用性保护、现状保留、规划新建等五种策略。绝对保护：还原日本侵华时期的历史原貌，佐证感受历史罪证。保护性利用：根据现状保护建筑室内空间格局、内部结构及建筑材质，引入与之相匹配的功能业态。形成1层商业、2层办公、3层居住的功能模式。利用性保护：利用现状破坏损毁程度较为严重的保护建筑，依据新型城市功能业态对空间的需求，可一定程度上进行内部空间改造，如小型办公区及酒店。现状保留：根据现状城市生活区内的建筑拆改难度及对于城市配套功能的承载，予以保留现状。规划新建：为了满足城市发展新时代对于产业功能及城市配套的更高要求，安置腾迁建筑，新建地面建筑，拓展地下空间，如展馆、地上小体量商业建筑、安置建筑等。

（四）动态保护与利用体系

依据项目特征需求，本着保护优先的基本原则，为了促进项目良性有序的发展，分析腾迁安置、城市形象提升、资金平衡等各方面因素，建议本次规划实施分为三个步骤。一期：项目启动区，也是本次规划设计任务的核心区。以保护历史建筑为首要任务，营造历史游览空间，升级植入匹配于时代发展、城市要求、游览配套的服务功能，对部分保护建筑实施保护性利用，在现状1层商业的基础上引入2层办公及3层公寓功能；利用三联楼的独特空间区域，利用性保护建筑，打造主题酒店；腾迁友协商场及5层居住区，建设以展示731历史、平房区沿革及工业发展历程为主题的展馆，周边规划项目核心广场；充分利用地下空间，打造平房核心区地下商业街；依托公社大街打造商业景观步行街，形成核心轴线空间。建立初步名片形象。二期：整治提升基地内部城市生活区整体环境，形成符合城市发展要求、生活惬意、秩序井然的城市功能空间。进一步提升项目的整体环境品质。三期：以保护性利用的方式发展剩余保护建筑。打造微小型办公区、文化创意产业区及主题公寓区等三大功能性保护空间，最终形成文物保护与城市发展的和谐共生。

时间快步走过了70年，但那段刻骨永久的悲愤之情并未随着时间的消逝而慢慢变淡。战争是对整个中华民族精神的无情践踏，我们无时无刻不在回望历史，我们无时无刻不对战争充满愤恨。战后至今，日本无耻的历史态度又无时无刻不在伤害我们的心灵，践踏我们的尊严。谨记那段历史，奋发于现在，展望美好未来，发展一个强大的中华民族，实现美丽的中国梦，是对中华民族屈辱史的最好回应。

项目信息及版权所有

业主：哈尔滨市规划局
建设地点：黑龙江省哈尔滨市
规划设计：北京建院约翰马丁国际建筑设计有限公司
设计团队：1.段毅 2.罗健敏 3.张达志 4.朱颖 5.乔崎 6.韩涛 7.邢屹 8.李莉 9.张薇 10.王充 11.陈瑜
规划面积：19.8公顷
设计时间：2015年
获奖情况：
侵华日军第731部队士官楼遗址保护与利用修详规划竞赛一等奖
摄影/图片版权：北京建院约翰马丁国际建筑设计有限公司
撰文：张达志

柒记 Note Seven

文化产业实践
Cultural Industry Practice

海鹊落原创音乐产业基地概念规划设计
Conceptual Plan and Urban Design for Haijingluo Original Music Industry Base

海鹊落原创音乐产业基地，通过对于中国“泛音乐”时代的音乐产业现状特征分析，
加之通过对美国百老汇、韩国的经验借鉴，找到了“音乐+”的产业格局新方向。
海鹊落原创音乐产业基地规划方案以海鹊落历史文化和音乐主题文化为核心构思原点。
隐喻“万鹰之神”海东青鸟的三片羽毛为规划立意元素，
规划李德伦音乐厅、实验剧场、音乐魔幻城三大建筑核心；
提取音乐主题的最美曲线，规划形成串联三大核心的音乐产业功能轴线，
形成一轴串三心、三心带三区的总体空间架构，
最终将它打造成具有交通体系多样便捷、绿色景观体系融合、科技与文化互动、
产业与城市共荣等丰富功能的可持续发展的城市空间。

While analyzing the status quo of the Chinese music industry in the Pan-music era and learning from the experience of the US Broadway and South Korea, the base planners point out a new development direction for the music industry, which is "Music +". Its design plan is centered on the local history and music-themed culture. Inspired by the shape of three bird feathers, Li Delun Music Hall, Experimental Theater and Music Magic Town are designed as three building compounds at the center of the base. And a beautifully-designed axis of music functions links the three compounds together. As a result, an overall spatial layout takes shape, where one axis strings three centers and three centers drive three zones. Ultimately, the base will become a sustainable urban space that features convenient traffic, green landscape, harmonious interactions between technology and culture, and co-development of industries and city.

海鹊落原创音乐产业基地位于北京市昌平区北七家镇，由奥体中心沿中轴向北，立汤路东侧。总规划用地75.36公顷。

通过对于中国“泛音乐”时代的音乐产业现状特征分析，以及相应利好政策的不断推出，加之通过在美国百老汇、韩国的经验借鉴，我们找到了“音乐+”的产业格局新方向。为了促动推进中国原创音乐的发展，我们急需建立配套完善的以创意、实践为主的实验田。为中国的音乐在国际舞台崛起，我们愿共同打造梦想的家园。

海鹊落原创音乐产业基地规划方案以海鹊落历史文化和音乐主题文化为核心构思原点。隐喻“万鹰之神”海东青鸟的三片羽毛为规划立意元素，规划李德伦音乐厅、实验剧场、音乐魔幻城三大建筑核心；提取音乐主题的最美曲线，规划形成串联三大核心的音乐产业功能轴线，形成一轴串三心、三心带三区的总体空间架构，最终将海鹊落打造形成交通体系多样便捷、绿色景观体系融合、科技与文化互动、产业与城市共荣、职住均衡、可持续发展的城市空间。

一、产业策划——“音乐+”新模式

随着互联网、新媒体技术的不断成熟，用户消费群体的更迭加速，音乐产业的形态和模式都发生了巨大变化，中共中央办公厅、国务院办公厅印发《国家“十三五”时期文化发展改革规划纲要》，明确将“音乐产业发展”列入“重大文化产业工程”中，从国家政策层面对音乐产业释放出又一重大利好。但是当前我国的音乐产业尚未形成成熟的商业模式，必须要通过加强人才培养、产业评估、版权保护等措施

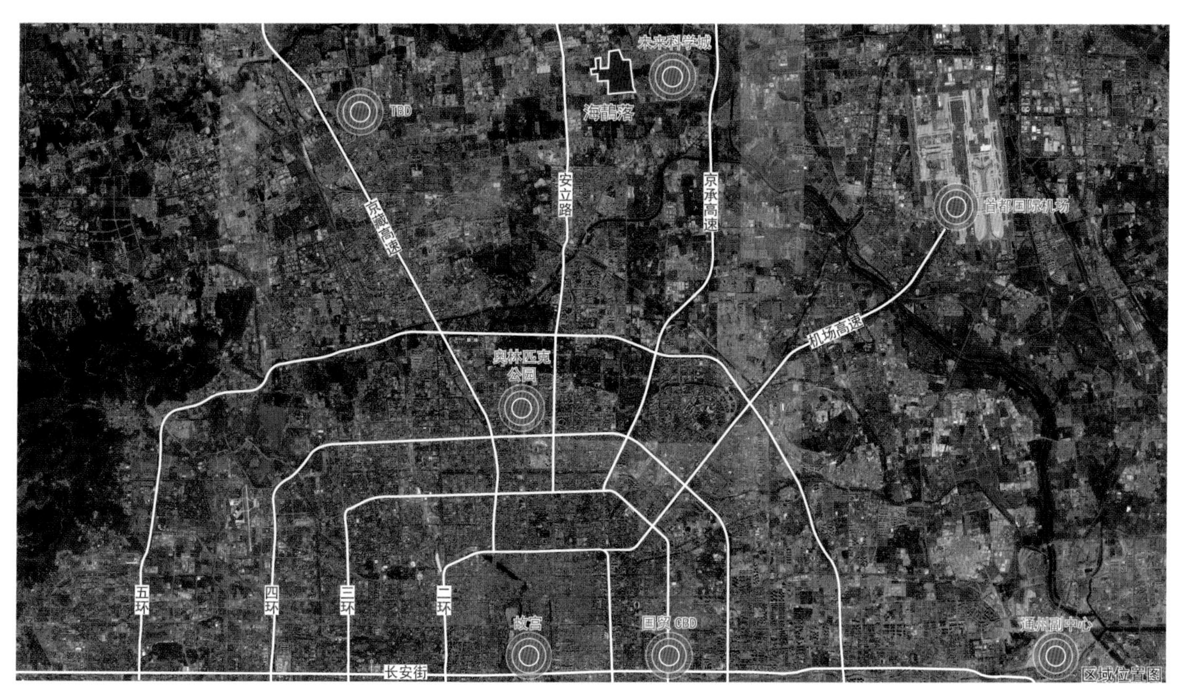

图 7-1 区域位置图

来推进产业升级发展。

北京作为国家首都、世界城市、文化名城和宜居城市，我们的产业定位必须契合站在文化发展战略和经济发展战略的高度，立足首都、重视本土、辐射全国、观照国际，高标准明确产业定位，全力打造具有首都优势、本土特色和国际竞争力的产业，形成具有广泛影响力、辐射力和可持续发展的强势产业，在推动首都的文化经济发展的同时，有力地推动首都的文化建设和发展。

（一）时代的召唤——“泛音乐时代来临，千亿市场即将腾飞”

1.泛音乐时代的核心——呼唤优质的音乐作品

近两年来，我国音乐行业发展多点开花：新生代偶像的崛起及人气屡创记录，互联网演艺平台增速迅猛，女主播月收入达到百万级。我们认为，音乐行业商业模式日趋多元化，音乐的商业价值开发也更加深度化，我国音乐发展进入新篇章，“泛音乐”时代来临。

泛音乐时代的到来标志着我国音乐行业将摆脱传统的单一盈利模式，内生增长和向外拓展将使得我国音乐行业进入多点开花的局面。内生增长方面，正版音乐时代，音乐作品销售将回到正轨；艺人形象的价值将因粉丝经济效应成倍放大。向外拓展方面，粉丝俱乐部、粉丝周边产品等庞大经济潜力将得到开发；互联网演艺平台的快速发展，给行业发展注入新动力；IP开发体系不断完善，音乐作为极佳的IP 资源的价值得到充分挖掘。我们认为，内生增长和向外拓展促音乐行业商业模式的升级，优质的原创音乐作品，将助力我国音乐产业的腾飞。

2.中国音乐产业的春天——2015年中国音乐产业突破3000亿大关

（1）2015我国数字音乐元年

“剑网2015”重点整治网络音乐版权，标志着我国正版音乐时代到来，付费模式即将形成，从全方位利好音乐产业的发展。

有别于国际惯例，中国网民长期习惯免费下载音乐、听音乐，需要过渡的时间更长。但是我国网民移动支付习惯已经大为改善，一旦行业竞争格局明朗，再辅以版权监管和数字专辑销售等创新措施，用户的付费习惯一旦形成数字音乐平台将拥有雄厚的资金、丰富的版权资源。

以实体唱片为代表的传统音乐产业规模进一步下滑，而音乐演出和数字音乐在技术和资本的双重影响下稳中有增，音乐版权与管理则受益于政策环境优化与结构调整，实现两位数以上的增长。中国音乐产业将在“十三五”期间结构性调整、融合共生的趋势下保持稳定发展的动能。

（2）国家相关政策

2015年国家版权局、文化部都出台了相关版权管理政策，国家新闻出版广电总局特别提出“打造上下游相互呼应、各环节要素相互支撑的音乐产业综合体系”；2017年中共中央办公厅、国务院办公厅发布《国家“十三五”时期文化发展改革规划纲要》，明确将“音乐产业发展”列入到“重大文化产业工程”中，“十三五”末文化产业成为国民经济支柱性产业。

（3）北京市相关政策

2016年北京市发改委出台《北京市文化创意产业发展指导目录》，将“内容的创作、制作、出版、传播”作为音乐产业发展的核心方向。2017年8月18日，北京推进全国文化中心建设领导小组第一次会议召开，提出要把首都文化优势转化为首都发展优势。国家文化产业创新试验区要进一步向文化创意产业引领区拓展。

（4）新形势下的新生态——“音乐+”

以优质原创音乐内容为核心，以突破产业边界为路径，以分享经济为纽带的“音乐+”发展模式成为音乐产业发展的新生态：音乐+互联网、音乐+现代金融、音乐+高科技……

我们国家的音乐元素不断出现在国际舞台上，300年前由意大利作曲家普契尼创作的歌剧《图兰朵》，讲述中国元朝公主的故事；我国的古典乐器，如二胡、古筝、琵琶等以及中国的诗词都频频登上了国际舞台。中国不缺优质的创作元素，但至今中国的原创剧目并没有登上百老汇的核心舞台，中国的音乐作品至今未在格莱美获奖，我们是否可以思考一下？

（5）我们的思考

不缺市场需求，缺优质内容；不缺人才储备，缺高端人才的培养机制；不缺演出场所，缺生产实践工作室；不缺资金投入，缺成规模的产业；不缺政策支持，缺产业的配套机制。

（6）经验借鉴

百老汇，它分为内百老汇、外百老汇及外外百老汇，但这三部分并不仅仅是非空间布局的划分，而是随产业发展规律自然而然形成的。内百老汇的座位是600人以上，能在内百老汇上映一定是灯光、音响等所有配置全面的的剧目，如果上映单独剧目风险较大，而外百老汇及外外百老汇恰好为内百老汇做了补充。百老汇的结算方式与我国相比，他们是要把一部戏从种子长到参天大树，从桌读到台读，观众从几十人到上百人，不断根据观众的反映进行修改，而这样的剧目才能真正经得起考验。他们是在版权、版税、衍生品及全球的巡演来长久获利。而我国是用一个创意换一笔钱，省下的部分就是利润。这就是为什么这么多年来中国缺乏优质内容的原因。

除了外部的空间支持产业的发展，百老汇内部还有许多成功之处：集群化的剧场分布、专业化的资源配置、市场化的剧目风险投资、创作主题的多样化、法制化的产业运作、巡演与衍生品及结盟非营利机构。

韩国在1997年遭遇金融危机，他们发现文化的前期投入较小但产值较高且文化与其他行业的融合度较高，于是基于韩国自身提出了4点发展模式：①演艺界多元化发展；②传统文化创新发展；③演艺的扩散性；④集群演艺方面。而最主要的是为文化产业量身定做的政策以及相关法律的修订等。

百老汇的成功之处在于艺术创作规律，而韩国是在政策推动下文化产业的大发展。

图 7-2 百老汇《猫》音乐剧（组图）

图 7-3《歌剧魅影》音乐剧（组图）

美国在1978年人均GDP超10000美元，成为文化产业快速发展期；韩国在1998—2003年人均GDP超过10000美元，也成为文化产业快速发展期；2016年我国人均GDP达8000美元以上，预计在未来几年超10000美元。人均GDP超过10000美元意味着娱乐文化消费进入爆发期。

（7）总结

“泛音乐”时代带来产业机遇。

“音乐+” 改变产业地缘结构。

我们具备世界一流的音乐展示平台，但缺乏具有世界影响力的音乐作品。我们急需建立配套完善的以创意、实践为主的实验田。

（二）文化自信—— 事关一个民族精气神的凝聚

中国交响乐奠基人李德伦大师的长女李鹿老师说，“我是子承父业。从20世纪20年代开始，肖友梅、贺绿汀先生在我国进行音乐的普及。我父亲从1980—1987年共在20余城市举办音乐讲座，每年平均要讲50次左右……”李德伦大师“游说”三届文化部长建起了北京音乐厅，让更多的普通老百姓走进音乐厅。对我们来说他不仅是中国交响音乐的奠基人，更是为音乐的未来之路指明了方向，鼓舞了中华民族自信。

谭利华老师放弃了英国的录取通知书，毅然留在了中国交响乐的舞台上，当谭老师和我们分享委内瑞拉青少年交响乐团的培养模式，也想在国内创办青少年交响乐团，这何尝不是在走李德伦大师走过的道路？还有程进老师1993年从美国百老汇回到中国，他说：“美国的音乐元素很丰富，但那不属于我，我要回中国。”回国后他推出了郑钧、许巍、田震、韩红等一批重量级的歌手。我们有很多这样的老师，他们怀着赤子之心、报国之梦，愿为中国的音乐在国际舞台崛起聚沙成塔、 乘风破浪。

（三）项目介绍——打造共同的梦想家园

1. 项目定位

汇聚高端音乐产业资源，深度融合跨界要素，打

图 7-4 实验剧场效果图

造音乐产业集聚区。

表7-1 产业业态权重表

业态分布 专业板块		A 创作与研发	B 教育与培训	C 版权与交易	D 展览与演出
1	古典	1A	1B	1C	1D
		1	5	2	4
2	流行	2A	2B	2C	2D
		5	4	5	4
3	音乐剧	3A	3B	3C	3D
		5	4	5	4
4	配套支撑	4A	4B	4C	4D
		4	2	1	5

注：权重：0~5级；1A~4D中权重最大的几个就是园区重点发展的产业内容。

表7-2 产业业态内容表

演出产业孵化器	流行音乐	音乐剧	音乐未来城
演出院线 教育院线 衍生业态	唱作人培养 音乐创作 音乐制作 艺人经纪 版权管理	原创剧目 实验剧目 人才培育 版权版税 驻场演出	音乐魔幻城 高端设备与艺术展

注：四大产业内容有：①演出产业孵化器；②流行音乐；③音乐剧；④衍生业

演出产业孵化器，将建立一个以李德伦命名的音乐厅（容纳800人）、多功能琴房170间及20个网络直播间。演出院线以驻场演出为主，不仅为当地打

图 7-5 项目构思生成图（组图）

造城市形象名片，也通过剧场演出不断修改剧目，打造优质精品剧目。配套网络直播室将对接全球30多个国家，实现优质节目的走出去，将国外好的资源引进来。同时联合国外知名经纪人团队，改变中国传统“保姆制”经纪模式，对基地的艺人进行包装打造。通过广告宣传，基地每年上百场的优质演出及明星效应能很好地带动广告媒体的发展，带来不菲的经济效益。教育院线涉及钢琴、声乐、交响乐等高端教育培训，我们的师资团队将在后面为大家呈现。

流行音乐，主要是唱作人的培养，音乐的创作、制作，艺人经纪，版权管理，园区的版权统一管理，建立发布、展示、交流的平台。

音乐剧，集创作、表演、经营、配套管理为一体。设有30个黑匣子剧场，主要负责音乐剧的排练，晚上可对外开放，通过观众反映不断进行剧目的修改。引进国内外50家音乐剧工作室，负责剧目的创作、制作、推广工作。我们要使园区每一个工作室都成为独立制作商，另外园区内建立演职人员协会，实行统一管理。

音乐未来城，含音乐魔幻城与高端设备及艺术展。音乐魔幻城以音乐为主题，与科技融合，通过“玩音乐”增强趣味性，引发“看音乐”的欲望，感受“学音乐”的快乐。高端设备及艺术展以世界各地音乐为主题，打造“多元化、多样性”特色展馆。分为两个区，高端设备展与艺术展。高端设备展将集合全球最高端的音乐设备、灯光设备、音响设备、舞美设备等。艺术展第一部分为非遗展区，通过文化遗产的各种实践、表演、表现形式、知识体系等，增强世界各国对中国的认同感和历史感，从而促进了文化多样性，更激发人类的创造力。

我们期待这一幕在中国音乐产业中出现，成为展示中国文化自信的舞台！

二、空间规划——鹊落吉地、乐起未来

泛音乐时代的到来，为音乐产业发展之路提出

音乐原创主题区
❶ 李德伦音乐厅
琴房及配套
网络直播室
❷ 录音棚
❸ 创作中心 & 黑匣子
❹ 孵化器 & 商业
❺ 创意工作室
❻ 音乐主题酒店
音乐剧原创主题区
① 实验剧场
② 网络直播室
③ 录音棚
④ 创作中心 & 黑匣子
⑤ 孵化器 & 商业
⑥ 创意工作室 & 商业
海鹊落公园
① 海鹊落公园主入口
② 海鹊落公园次入口
③ 海鹊音乐广场
④ 水上音乐喷泉
⑤ 露天秀场
⑥ 音乐纪念园
音乐未来城
❶ 音乐魔幻城
❷ 高端设备及艺术展馆
❸ 音乐科技孵化中心

图 7-6 功能布局总平面

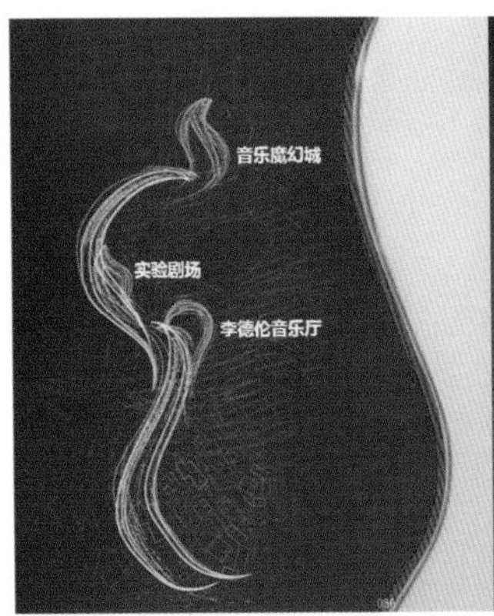

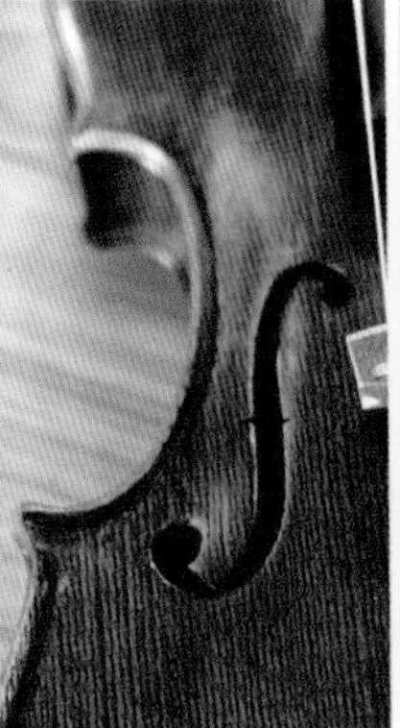
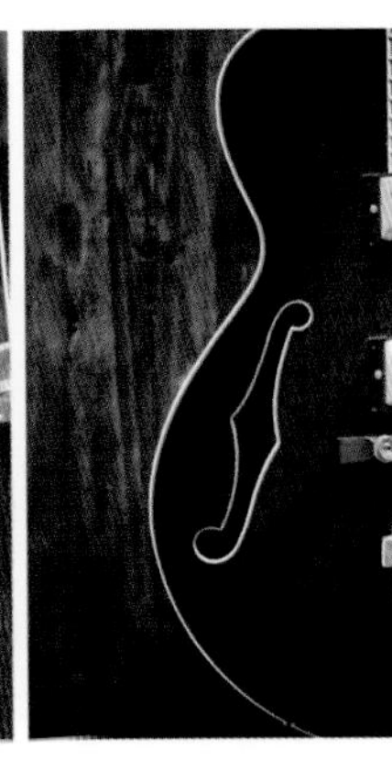

图 7-7 项目设计概念生成图

了新的产业发展高度及广度要求，为了适应产业发展的时代要求，引领音乐产业的发展。我们创造了“音乐+”的产业发展运营模式，为中国的音乐产业发展提供了新思路，为北京的文化之都建设注入了新活力。

那么，“音乐+”全新的产业模式将落户何处？又应以怎样的产业空间承载构建？与当地历史文化将如何融合传承发扬？如何创造一个益于音乐人创作、交流、展演的城市产业空间，一个适于全民音乐主题的城市休闲空间？一个利于发展的城市服务功能空间？一个有助于生态共融的城市生态空间？这是我们的核心规划设计议题，同时更是助力音乐产业发展、昌平城市发展更新、北京文化建设的行为和举措。

（一）地利承载——音乐产业基地为何落位海鹊落

1.全国视角——全国中心、音乐城市

北京是全国的文化中心，也是国内乃至国际文化人才集聚的高地。从全国的音乐产业发展格局来看，根据国家广电总局对于音乐产业核心城市的定位，北京、上海、成都、广州4个城市为国家音乐产业基地城市。北京作为我国的首都，是我国音乐产业的集聚中心，也是亚洲音乐创作和音乐人才的汇聚之地。

2.北京视角——文化交流、北京引领

2017年9月27日，中共中央国务院批复了《北京市总体规划》，北京市新总规将北京定位为政治中心、文化中心、国际交往中心及科技创新4个中心，音乐产业符合北京构建文化中心的定位要求，同时音乐产业也是国际交往的重要媒介。

3.昌平视角——文化协同、昌平响应

在京津冀一体化、北京产业和人口疏解外溢的情况下，昌平、顺义、通州及大兴共同形成了北京市功能协同圈层，其中昌平在区域产业功能发展方向和文化产业空间承载适宜性都具备较好的产业条件和环境基础。

4.海鹊落视角——文化承接、落位海鹊

《北京市文化创意产业功能区建设发展规划（2014—2020年）》提出了构建“一核、一带、两轴、多中心”的空间发展格局，其中在北京北部构建未来文化城功能区，与北京城市轴线历史文化保护区共同形成北京文化服务产业轴，是规划提出的北京文化产业发展格局的重要构成。在该规划中，未来文化城功能区考虑位于北京北中轴附近，昌平区南部。海鹊落的位置与《北京市文化创意产业功能区建设发展规划（2014—2020年）》中建议的昌平文化产业布局相符合，是北京文化服务产业轴北端节点。

①北京轴线、现代演绎。在北京建城史上，元、明、清三朝的更迭演进构建了中轴线的传统历史文化区域，自新千年以来，尤以2008年奥运会的举办，北京中轴线向北延展，形成了以体育文化为特征的新时

代节点，海鹊落原创音乐产业基地项目将以音乐主题文化为特征，为中轴线的现代文化演绎增添浓重的一笔。

②科技文化、双创发展。海鹊落同时位于规划建设中的“京北产业带”上，其西接北京科技商务区（TBD），东临未来科学城及首都机场临空经济区。海鹊落原创音乐产业基地的规划发展将触媒带动形成京北产业带科技创新、文化创意的双创发展的新格局。

③唇齿相依、互动共荣。未来科学城是北京“三城一区”中重要的科技创新城，是北京4个中心定位中科技创新中心建设的主平台之一。海鹊落位于规划的未来文化城之中，紧邻未来科学城。未来文化与未来科技在此地互促发展、在空间上唇齿相依，在产业功能上互动共荣。

④文化科技、交往融合。法国著名文学家福楼拜有句名言：“越往前走，艺术越要科学化，科学也要艺术化，两者在山麓分手，又在顶峰聚集。”这句话精辟地阐述了科技与文化两者互相促进发展的关系。

总结来看，海鹊落南接北京中轴线，是北京文

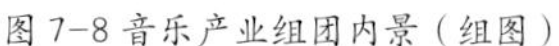
图 7-8 音乐产业组团内景（组图）

图 7-9 东北方向鸟瞰图

图 7-10 音乐未来城效果图

图 7-11 李德伦音乐厅效果图

图 7-12 音乐产业组团入口效果图

图 7-13 大师工作室区鸟瞰图

化服务产业轴的北端节点，东西紧邻TBD及未来科学城，东北方以雁西湖为核心的国际交往中心，是文化、科技、国际交往的聚焦融合之地。

⑤音乐文化产业集聚。通过对北京音乐产业及相关文化产业的调研，发现与之相关的产业功能目前基本聚集于朝阳北部、海淀东部、昌平南部及首都机场周边。集聚的相关产业及从业人才为海鹊落的未来奠定了强大的音乐产业发展基础，是音乐及相关产业外溢的良好承载地。

⑥四通八达、交通便捷。海鹊落基地至机场、高速及城区均在半个小时车程以内，区域交通联系便捷。基地周边现已形成多条城市快速路及城市干道，未来地铁5号线的北延计划及17号线的规划建设也为基地的发展提供了良好的公共交通保障。

⑦生态海鹊、创意融合。即将建设的海鹊落公园的良好生态环境，能为音乐人提供惬意的创作空间。

总的来说，我们认为：北京是全国文化交流中心，也是音乐产业集聚之都；音乐产业符合北京文化中心和国际交往中心的定位要求；昌平是北京东北部音乐产业积聚、外溢的理想承载地；海鹊落位于北京北部产业带和文化产业服务轴交会之处，具备激发创作灵感的优美环境。

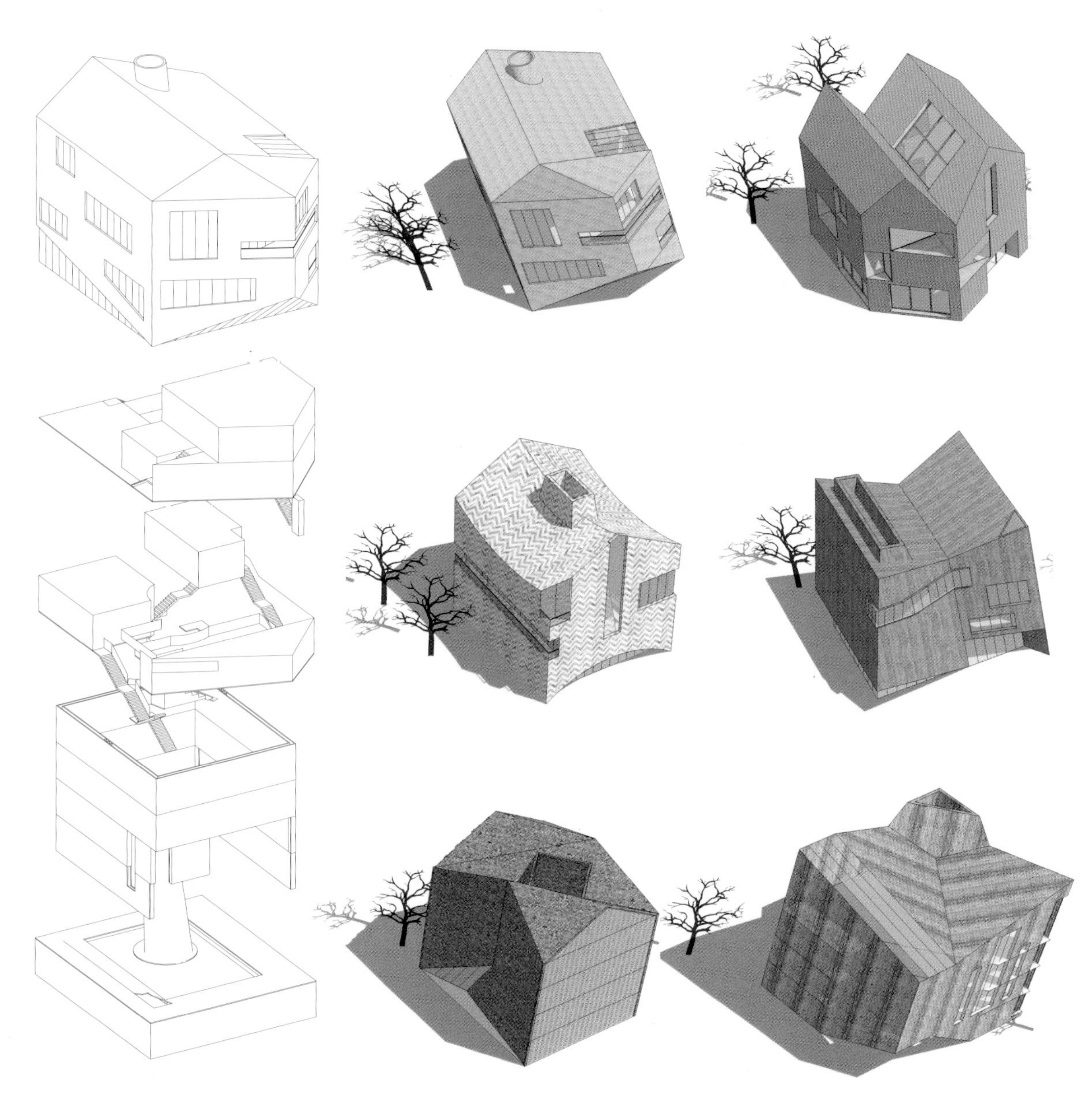

图 7-14 大师工作室单体空间拆解图　　图 7-15 大师工作室多元可定制化单体图

（二）空间规划——如何打造适合音乐产业发展的城市空间

1.规划引导

海鹊落原创产业基地初步选取了北七家镇定泗路以南，北清路以北，海鹊落公园西南，七星路两侧的75.36公顷用地，依据现有的《昌平新城规划》及《北七家镇实施方案》，该用地中建设用地为52.4公顷，总建设规模为88万平方米。

2.方案构思

音乐是无国界的语言。贝多芬说："音乐，是比一切智慧、一切哲学更高的启示。"

海鹊又名海东青，被称为"万鹰之王"，相传十万只才能驯化成一只海东青。契丹国志中记载："五国之东接大海出名鹰"，即为海东青。康熙有诗对其赞誉称："羽虫三百有六十，神俊最数海东青"，体现了海东青的勇敢和智慧。

海鹊落原名"风火屯"，名称的由来有元朝的王爷围猎和明朝燕王朱棣赐名两种说法。两种孰是孰非虽无从考证，但海鹊鸟的祥瑞曾降临此地是毋庸置疑的。

图 7-16 项目推广 LOGO

同样，音乐也对方案设计给予的更高的启示，韵律灵动的音符得益于乐器的华美演绎。不同的乐器类型似乎都有着同样的特征，我们看到提琴、吉他、竖琴都有一条优美的曲线轮廓。

规划方案通过一条优美的曲线将海东青翱翔于海鹊落上空散落了的三片神羽毛幻化成的原创音乐产业

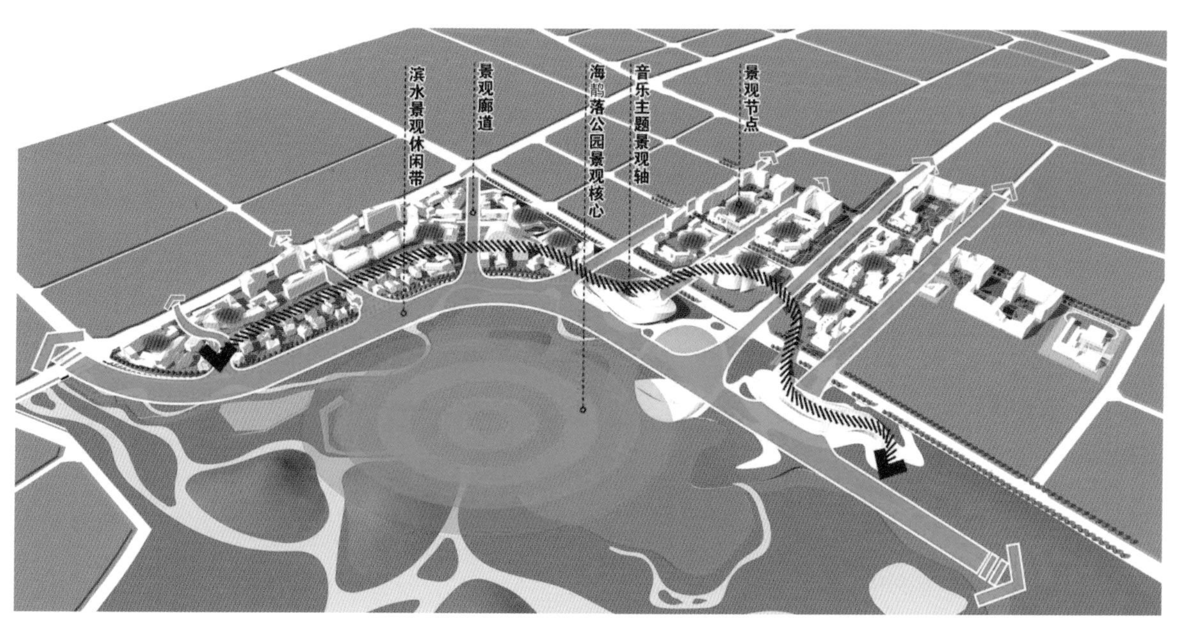

图 7-17 景观体系规划图

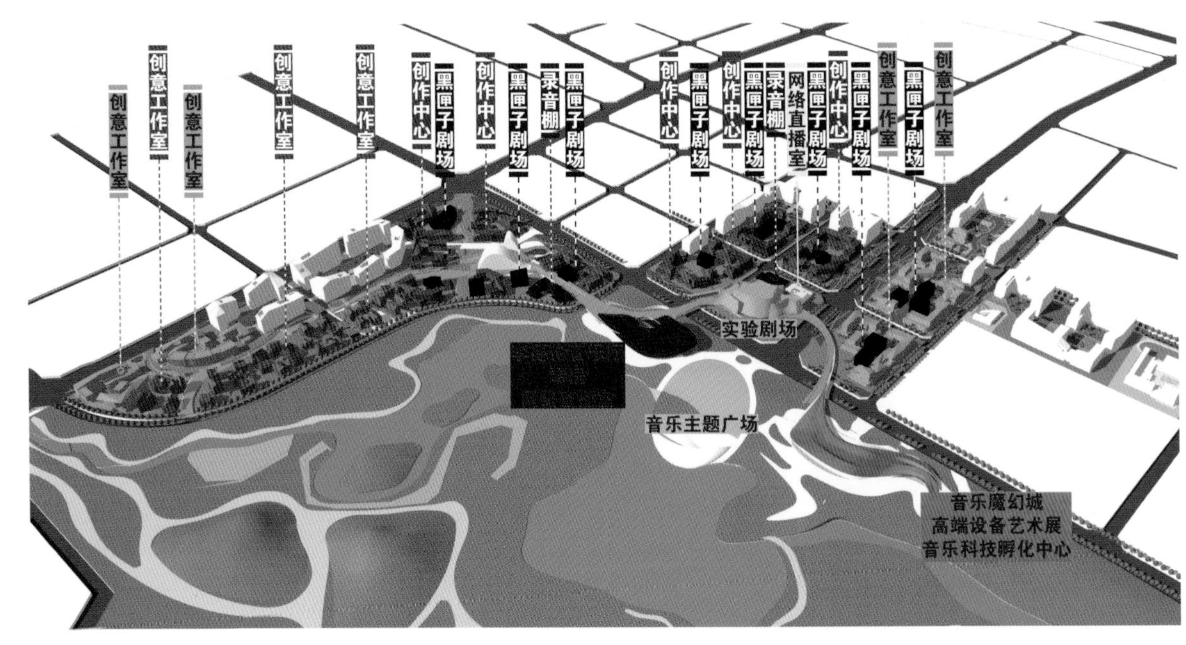

图 7-18 业态分布规划图

园中三处核心功能建筑（李德伦音乐厅、实验剧场、音乐魔幻城）连为整体，形成了规划方案的整体骨架。

3.空间构建

音乐产业功能轴线串联三大核心功能建筑，三个核心建筑引领三个音乐产业功能片区，一轴串三心，形成“乐颂华章、乐动海鹊、乐享世界”三个的音乐承载空间。

（1）第一个功能片区——“乐颂华章”

乐颂华章片区以西洋古典音乐、中国民乐及流行音乐的创作、演出、教育及版权交易为主题功能。

李德伦音乐厅是该区的核心引领建筑，除音乐厅之外，还包括配套琴房及网络直播间等。

创意工作室是为音乐行业名家名师提供创作、交流、生活的空间。

音乐主题街区是音乐创作、分享、交流、休闲的复合功能街区，有机立体的空间布局模式整合了录音棚、黑匣子、创作中心、孵化器、工作室、商业配套、酒店等功能，创造有活力的混合功能社区。

（2）第二个功能片区——“乐动海鹊”

乐动海鹊以音乐剧剧本创作、实践演出、交流互动、教育培训、版权交易为主题功能。

实验剧场位于该区的东北部，是实验演出和音乐剧产业的重要基地。

音乐剧主题街区是创作、实践、路演、休闲的复合功能街区，整合了黑匣子、创作中心、孵化器、工作室及休闲商业等产业功能。

在东西干道的北侧，布局了音乐科技中心。

（3）第三个功能片区——“乐享世界”

乐享世界以音乐体验、音乐休闲、音乐历史、音乐博览、艺术品及器材展览展示为主体功能，包括音乐魔幻城及配套的展览中心等。

三个功能片区通过在中央的三个核心建筑形成了“一园、一轴、三心、三区”的空间结构。

一园：海鹊落音乐主题公园。

一轴：音乐产业功能轴。

三心：三片羽毛构成的李德伦音乐厅、实验剧场、音乐魔幻城。

三区：音乐原创区、音乐剧原创及音乐科技区、音乐休闲区。

方案总建筑规模83万平方米。其中产业公建32万平方米，配套及居住51万平方米。配套及居住包括酒店3万平方米、商业6.1万平方米、保障性住房13万平方米、小学1.5万平方米、幼儿园0.4万平方米、住宅27万平方米。

4.体系支撑

（1）多样便捷的交通体系

机动车道路交通系统与上位规划保持一致，同时打造以音乐产业轴步行系统主骨架。

（2）蓝绿共融的绿色生态体系

打造以海鹊落音乐主题公园为核心，基地内多条绿廊向心渗透的绿化空间格局。

（3）错落有致的空间形态

园区最高建筑为60米，邻近公园为低层，逐渐向外围增高的梯度高度控制，形成环抱公园的天际空间形态。

在目前的概念规划基础上，将原定建设总量减少2万平方米，同时遵循上位规划道路交通体系路网结构，保证公共服务设施用地及绿地的总量要求。

（4）科学的城市建设理论支撑

园区将按照海绵城市、智慧城市、LEED绿色社区等城市建设标准打造绿色、安全、智慧、混合、可持续发展的音乐产业集聚功能区。

（三）未来愿景——海鹊落将创造一个怎样的音乐梦想之地？

我们希望这个音乐原创基地未来是音乐家、流行音乐人、音乐剧创作人创作、实验演出的沃土，是原创音乐市场推广的平台，是中国音乐走向世界的土壤，是一切喜欢音乐、爱音乐者的净土、是音乐娱乐的乐园，是音乐和艺术、科技结合走向巅峰的见证。

未来的海鹊落将是一个功能复合的音乐全产业生态圈；一个助音乐人成功的推动平台；一个易于创作、生态共融的自然环境；一个艺术交流、活力共享的音乐社区；一片孕育大师梯队、产城融合的沃土。历史上，祥瑞曾降临此地，如今我们以音乐的名义期许海鹊落的美好未来。

项目信息及版权所有

海鹊落原创音乐产业基地概念规划设计
业主：卓越集团
建设地点：北京市昌平区
城市设计：北京建院约翰马丁国际建筑设计有限公司
联合设计：首建品印国际建筑规划设计（北京）有限公司+北京工业大学建筑勘察设计院
产业策划：涌现集团
设计指导：李鹿
设计团队：1.朱颖 2.张达志 3.崔颖 4.刘玉锋 5.段毅 6.陈威 7.白雪 8.乔崎 9.王鹏 10.邢屹 11.韩旭 12.梁学强 13.林新业 14.王浩冉 15.柯谨 16.梁晶 17.张明哲
基地面积：75公顷
设计时间：2017年

撰文：朱颖 张达志 崔颖 刘玉锋 陈威

Unicare
北京优联美
2号柜台
1号柜台
Counter 1

捌记 Note Eight

医院改造实践
Hospital Renovation Practice

北京优联医院改造设计
Beijing Youlian Hospital Renovation Design

人性化设计是基于人体工程学的要求，基于对弱势群体的人类学调研，
实现在设计中先关怀人的个体。
通过提升建筑创作深度建构富有各种“表情”的城市医疗场所，
建筑师要从为人服务的主动精神入手，创作出不仅宜人且“像人类那样的东西”。
这并非是形式上的柔和触感，
而是需要适用、自然、美观、坚固的空间，真正从为不同人的角度，
实现“适用、经济、绿色、美观”的国家建筑方针，真实体现对病人及残障人的照料。

Personalized design needs to follow the human engineering principles. The findings of the research into the needs of vulnerable groups allow designers to put individuals at first place. On the basis, they can deepen their architectural design and create the urban medical facilities with various "looks". Starting from the initiative to serve people, architects are expected to design well-functioning and nice-looking buildings. By saying so, the author doesn't mean the soft touch in appearance but the practical, natural, good-looking, and concrete spaces. Truly around the needs of different people, their work will embody the national construction guideline for "practical, economical, eco-friendly and beautiful buildings", and show their wholehearted care for the patients and the disadvantaged.

该项目位于北京朝阳区东四环南路海渔广场内。林达海渔广场总占地7.9万平方米，总建筑面积近20万平方米，被定位为六位一体的中国首席海洋文化城市综合体，项目由1栋19层5A级写字楼，3栋15层酒店式公寓，1栋20层五星级酒店，2栋5层独立商业楼，1.5千米商业街及1.5万平方米大型地下超市和5万平方米以海洋文化为主题的海渔公园组成。

此医疗设计项目为利旧改造项目，为2栋5层商业办公、1栋5A写字楼和局部地下室进行高端医疗项目设计。原建筑使用功能为餐饮、办公的异形建筑，需改造为集急诊、内科、外科、儿科、妇科、皮肤科、口腔科、产科、眼科、耳科、鼻科、透析中心（60床阴性、8床阳性）、中医科、体检中心、高端健康管理及相应的三级标准检验科、放射科、万级手术室1间、千级手术室3间、千级负压手术室1间、百级手术室2间、血库2间、ICU、NICU、药剂科、供应室的260床二级综合医院。

北京优联医院是“大专科、小综合”，以JCI标准设计，以首都医科大学为依托，集合多个重点专科，引进国内外先进医疗设备，提供高品质、有尊严的综合性医疗保健服务。

拟改造建筑由地下商业、5层和19层商业写字楼组成，该建筑为框架剪力墙结构，使用年限为50年，建筑高度分别为24米、80米。

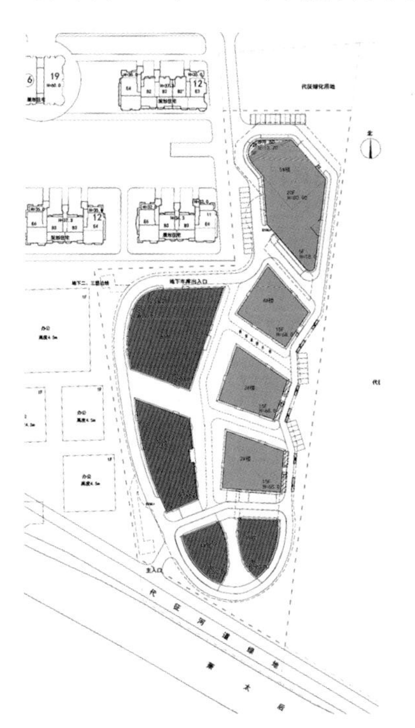
图 8-1 总平面图

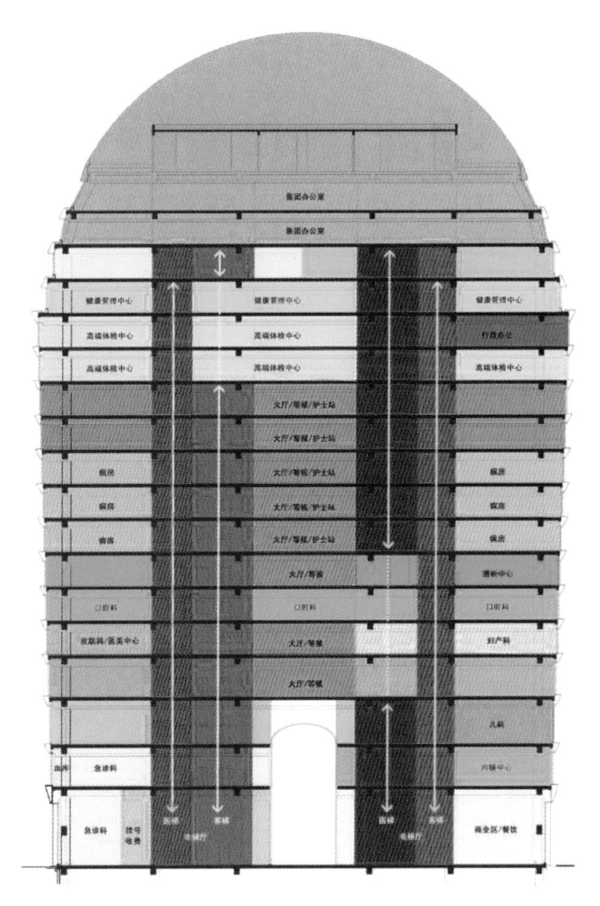

图 8-2 1# 楼功能剖面示意图

图 8-3 医疗展示区

一、项目实施

①项目实施的难点。既有保护建筑不管是改造为酒店、商业、办公还是医疗等功能性建筑，都面临如何解决保护与功能更新之间的矛盾。由于医院管理流程及相关的技术规范对医院的功能及流线有严格的要求，加之既有建筑的周边环境、结构体系、空间大小等方面限制条件的客观存在，改造设计面临更苛刻的要求。

②既有建筑改造的优点。通过对既有保护建筑的功能更新，让建筑发挥新功能，实现自身价值，激发城市活力，也为医院（特别是民营医院）的建设提供了更多的可能性。相对新建医院，既有建筑现成的建筑体量提供了容纳医疗功能的空间体系，可以缩短医院建设的周期。

③设计内部实施保障。项目组在保证安全前提下，对既有建筑实际情况进行调查研究、掌握详细的第一手资料。由于项目特殊性，图纸资料和现状情况不一致，为设计增加了难度，我们成立了建筑、结构、设备、电气、室内等各专业组成的项目组，在极短时间内快速完善方案。在设计过程中，项目组认真研究、学习、调研类似项目，总结经验，积累知识，及时解决技术和难点，在现场施工工程中，安排专员负责项目设计管理和协调，大大提高办事效率。

④外部设计协作。医疗项目复杂，涉及领域广泛，需要进行各种专业及厂商协调和合作。在与各方协作过程中，项目组组织多次协作单位召开项目会议，发现问题，解决问题，保质保量地完成从方案到施工图的全部设计工作，并且全部配合施工现场技术协调及竣工验收等工作。

二、设计策略

①保护和再利用。在既有办公楼改造中处理保护与功能更新的矛盾，采取的策略是在保障医疗功能前提下对建筑外观进行完整的安全性保护，以延续建筑时间和空间的记忆；对内部空间在尊重原有结构体系的前提下，将原办公功能完整地置换为医疗功能，实现办公楼的再利用。

②梳理和分析。设计须解决既有办公建筑性质改变为医院后结构抗震措施问题；解决建筑性质改变为医院后消防及相关规范的问题，如疏散距离、楼梯及通道宽度、电梯标准、卫生间设置、净高控制等；解决医院大型医疗设备荷载带来的结构加固问题；解决医院的功能设置和定位及各个功能区医疗流程问题；解决设备管线通道与结构体系之间的问题；解决建筑性质改变后土建设备的基本配置要求。此外，还须结合项目特点综合考虑：梳理用地周边环境，分析处理医院功能区与周边功能区关系的可能性；梳理医院各个出入口之间的关系，处理各条流线之间的问题；注意医院辅助用房，如污水处理、氧气站、污物存放等的位置。

③规划和保障。规划医疗流线是实现医院功能的核心，通过垂直交通体系作为交通主系统，构成水平和垂直的交通网络，经过对人员流线、物品流线、洁污流线进行详细的分类和梳理，做到使用便捷、互不干扰，并实现洁污分区和分流。

三、医院改造工程的重点

①建筑结构安全。结构措施是项目安全的支撑，既有建筑为办公或者商业建筑，医疗建筑人流大，各项系统使用频繁，医疗设备荷载较大，原有结构可能达不到要求，如核磁共振仪器等重大设备的加装，必须对原有结构进行系统检测并出具结构强度检测报告，根据检测结果进行结构加固。

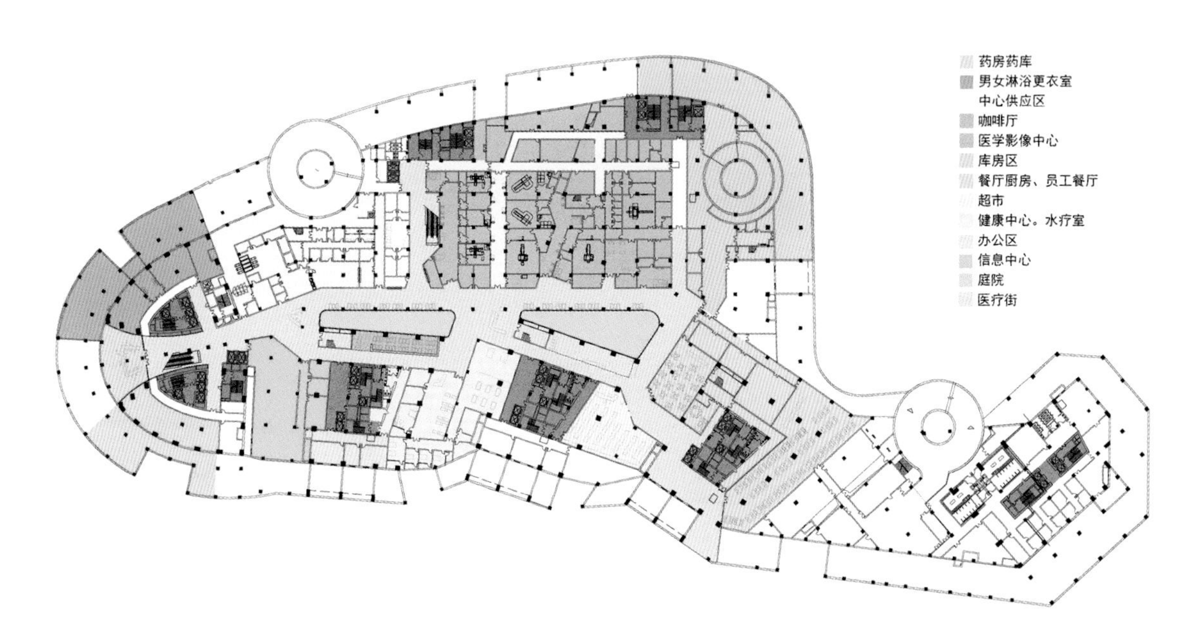

图 8-4 地下一层平面图

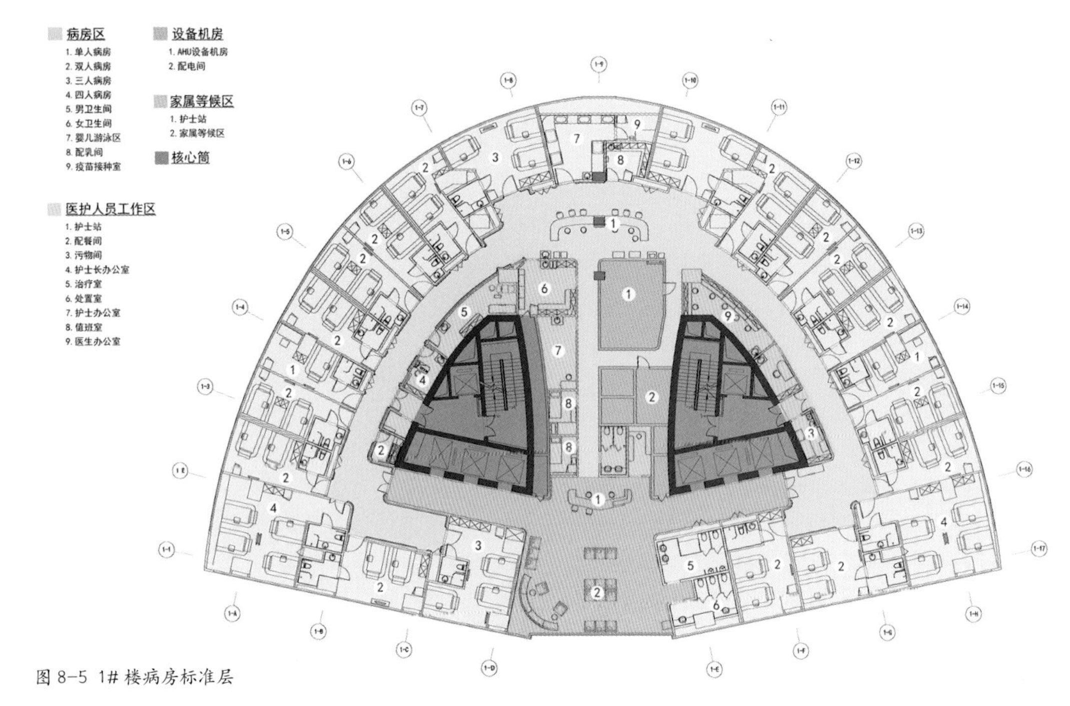

图 8-5 1# 楼病房标准层

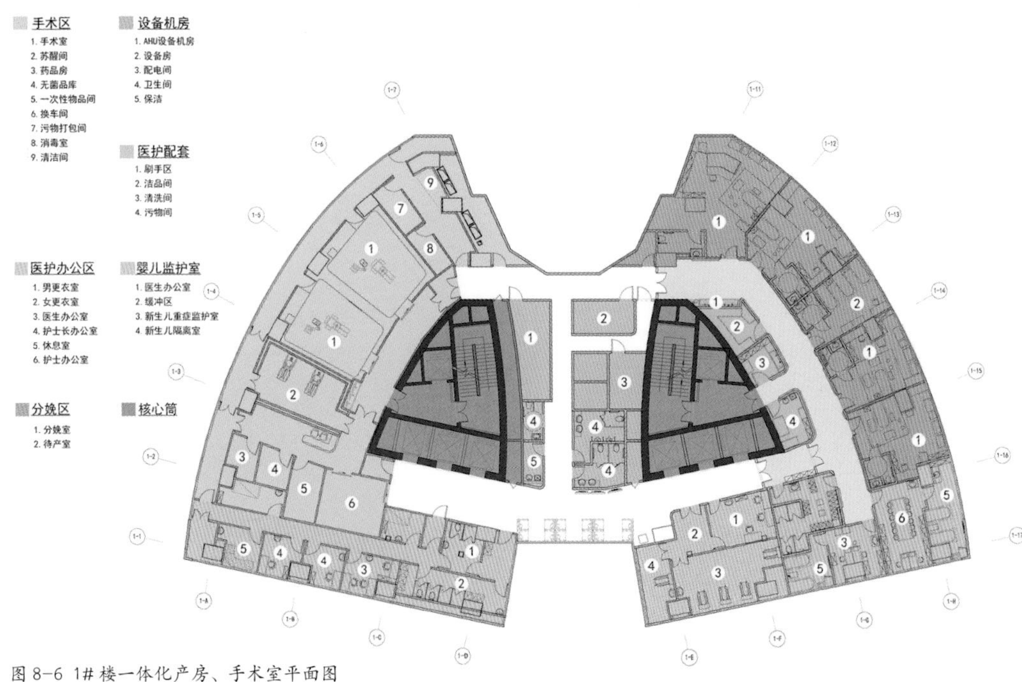

图 8-6 1# 楼一体化产房、手术室平面图

图 8-7 健康管理等候区

②医疗安全。从规划设计方案阶段就应该考虑医疗安全，这其中包括流线安全、物流安全、防辐射安全等。设计和基建人员应就以上问题深入与使用科室沟通，做到统筹安排。

四、项目改造的内容

（一）功能布局和流线

1.功能布局

由于医院设计的本身特有的复杂性和实施过程的的变化性，首先对医疗结构、功能和规模以及相关医疗流程、医疗设备、技术条件和参数，进行核对和审查，对医院各医疗功能单元之间的流程和医疗功能单元内部的流程进行设计和规划。

在该项目建筑功能布局的改造中必不可少的过程，就是和医院各科室进行交流，确定相关数据，其中有些数据不可或缺，如门诊诊室间数、急救抢救床数、各护理单元设置的病床数、手术室间数、重症监护病房（ICU）床数、心血机造影机台数、X光拍片机台数、肠胃透视机台数、胸部透视机台数、B超机台数、心血管彩超机台数、内窥镜台数。设计工作就是把这些医疗单元在现有的建筑空间中科学布局，优化既有建筑设计，更好地满足医护人员、就医人员的使用要求。

（1）1#楼医疗建筑主要功能

首层为商品展示厅、咖啡厅、挂号大厅、急诊科，2层为内镜中心和急诊科，3层为儿科，5层为妇产科、医美中心，6层为口腔科，7层为透析中心，8~12层为病房层，13~14层为高端体检中心，15~16层为健康管理中心，17~18层为集团办公区。

（2）7#楼医疗建筑主要功能

1层为挂号大厅、等候区、耳鼻喉科，2层为手术室，3~4层为高端病房，5层为办公区。

（3）6#楼医疗建筑主要功能

1层为多功能厅，3~4层为实验室，5层为培训教室。

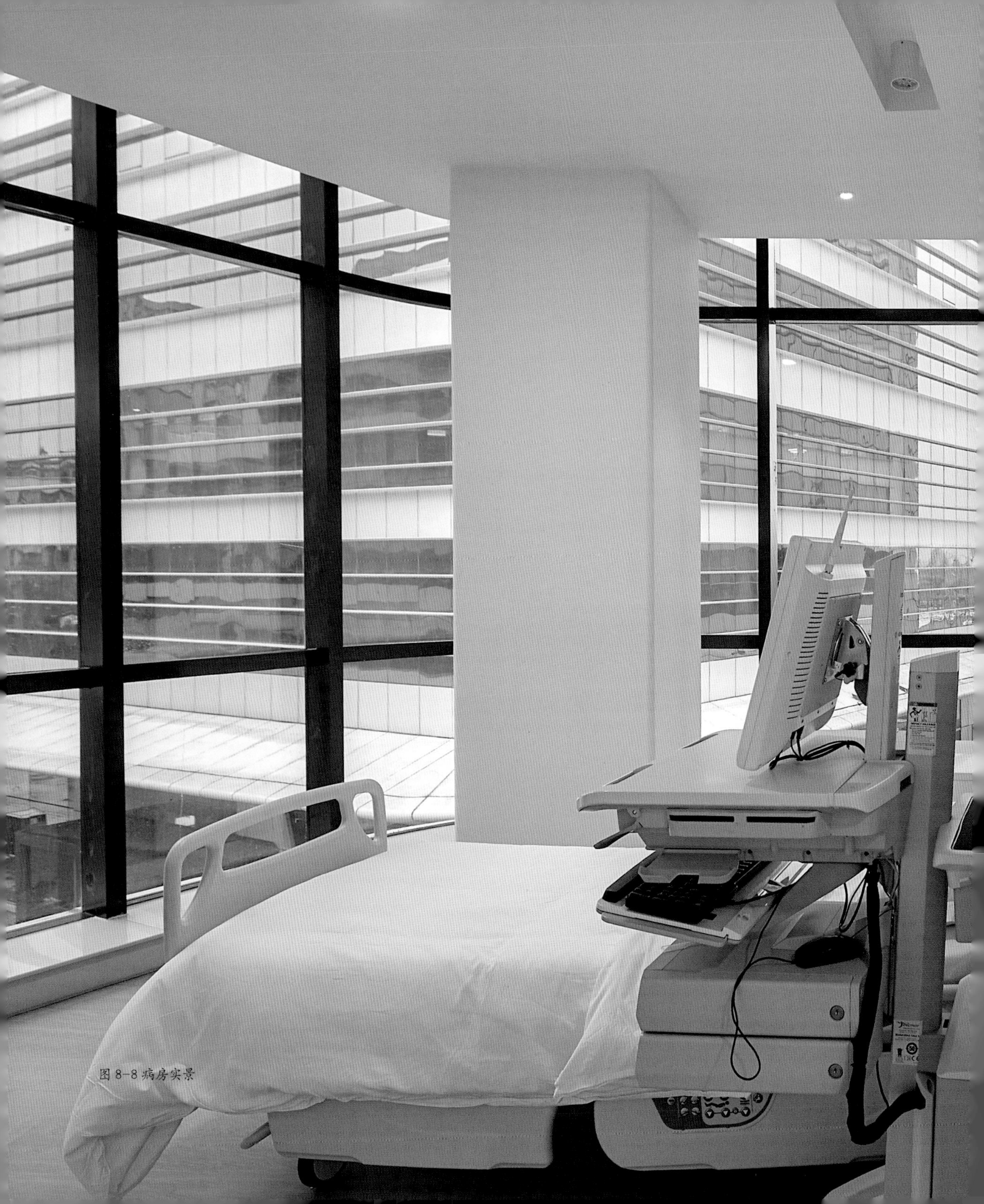

图 8-8 病房实景

（4）地下室主要功能

影像中心、信息中心、药房、检验科、厨房、餐厅。

2.医疗流线

梳理就医流线和医护人员的动线，高流量科室设置于底层，高私密性设置于高层，相关科室设置于同层或者相邻楼座，可灵活调节功能的设置，改造平面布局从而形成高效的空间，预测流线堵点，反复优化流线设计，直至流线呈最佳合理状态。

3.现代化的病房

根据现有面积及电梯运力，采用每层为一个护理单元的模式，提高了医院运行效率。标准层护士站居中设置，便于护士观察所有病房，缩短护理路线。充分考虑医院特殊性，合理分配电梯功能，高低分区，缓解了人流和物流，利用隐蔽电梯设置污梯，方便运输。由于建筑为曲面建筑形态，病房环绕东、南、西布置，视线良好的落地玻璃窗为每间病房提供良好的自然采光和通风。宽敞的病房内设置的挂衣橱、电视机、淋浴房等辅助设施，体现对患者的人文关怀。

（二）建筑室内设计

①构思。用充满设计感的室内空间打破传统医疗环境的机构形象，运用自然、环保、柔和的装饰材料创造更加舒适的室内氛围，平复就诊者的情绪。在人们踏入医院的这一刻来自设计方的心理治疗便已经开始。

②设计手法。将室内、室外与品牌的设计语言融合统一。弧形元素作为室内设计主要语言之一，其本身的二维造型也被三维化用于柜体、镂空等处贯穿始终。外立面富有韵律的造型元素被沿用于室内多处空间。

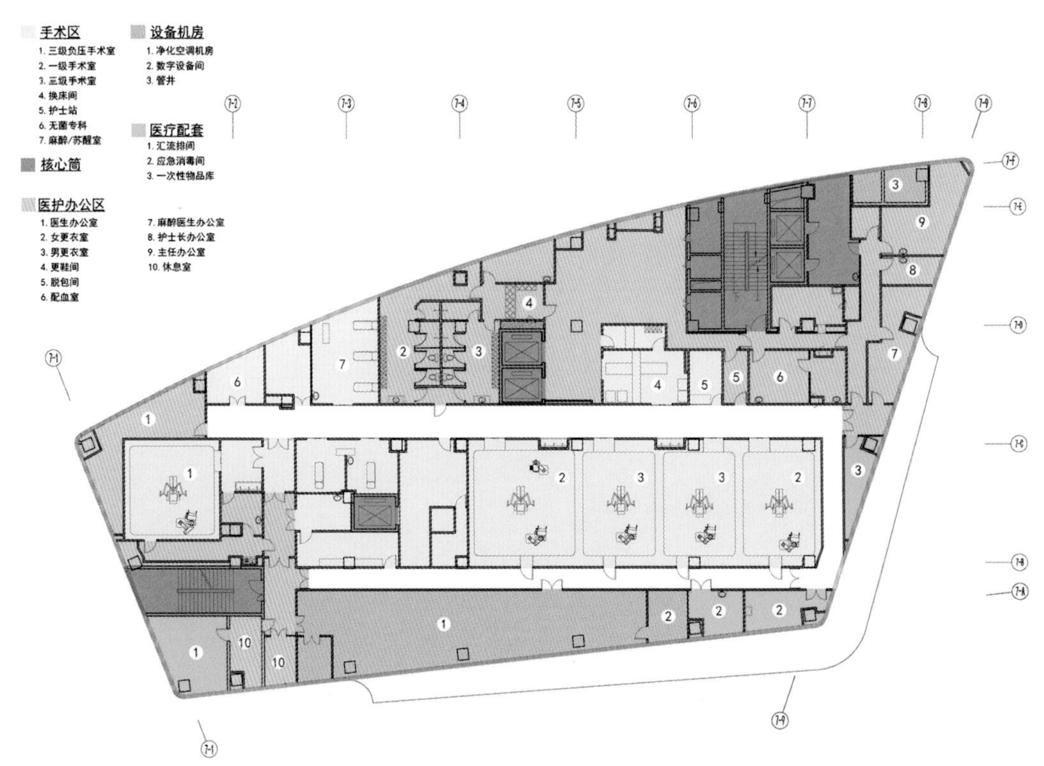

图 8-9 7# 楼手术室平面图

③突出医患空间的人文关怀。诊室、病房、候诊等患者经常使用空间均有良好的通风、采光，房间装修温馨、雅致，医院中的走廊、休息活动区尽量做到舒适、温馨；休息、购物、充电、餐饮等各种功能齐全、方便。

④注重建筑空间与医疗空间的契合。空间的大小、形状、尺寸以符合医疗流线的导向性需求为标准，而不是一味地追求高大。

⑤注重装饰材料的质感。隔墙材料和部分空间的装饰材料的使用特别注意选用色彩温馨、质感宜人的材料，如选用天然植被或带自然肌理的环保材料。

⑥重视光环境和色彩。在医院设计中，光和色关系密切。如果光的设计效果不好，不仅功能上难以满足患者及医护人员的使用要求，同时也难以发挥色彩的效果。灯光不仅是室内照明的需要，更是室内装饰的重要元素。考虑到医疗特殊功能，不同区域采用不同的照明形式，不同色彩和亮度的灯光、点光源、灯带、面光源的结合，满足了不同空间、不同导向性以及私密性的要求。

（三）建筑结构加固

本工程于2008年竣工，主体地上18层，地下3层，建筑高度83.80米，为全现浇钢筋混凝土框架—剪力墙结构，抗震设防烈度8度（0.2g）。原为商业、餐饮及办公使用，拟改造1~7层为医疗用房，8~16层为病房。抗震设防类别由丙类提高为乙类。

原结构在4层以上大范围的楼板采用现浇预应力空心楼盖技术，框架梁采用无黏结预应力梁。因各层隔墙数量增加较多，局部房间医疗设备较重，致使楼面荷载值比原设计有所增加，须针对楼板和梁进行适当加固。为避免伤及预应力钢筋而造成安全事故，减小加固施工中对原结构的破坏，设计人员针对不同构件采取多种形式的加固方法，如当梁的配筋相差较大时，采用粘贴钢板或加大截面加固法；当梁配筋相差较小时，采用粘贴碳纤维加固法；对局部楼板进行加固，采用以粘贴碳纤维为主、粘贴钢板为辅的加固法。

由于医疗建筑的特点，大量设备、电气管线穿过楼板和墙体，须新开楼板洞、墙洞数量较多。设计要求开洞应采用静力方法，开洞位置应躲开已有梁；开洞前应探明预应力钢筋准确位置，开洞时避免伤及预应力钢筋。新开板洞周边采用粘贴钢板方法进行加固；原有悬挑板上开大洞后，采用新加型钢挑梁、型钢边梁进行加固。

针对加固改造施工，设计方提出相关施工要求：①加固改造工程的施工必须由具有相应施工资质的专业公司完成；②构件进行加固前，应卸除原有隔墙荷载，楼板上不得集中堆放施工用料或建筑垃

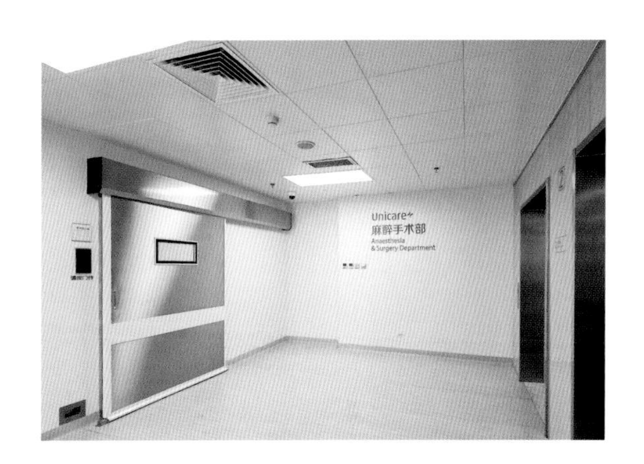

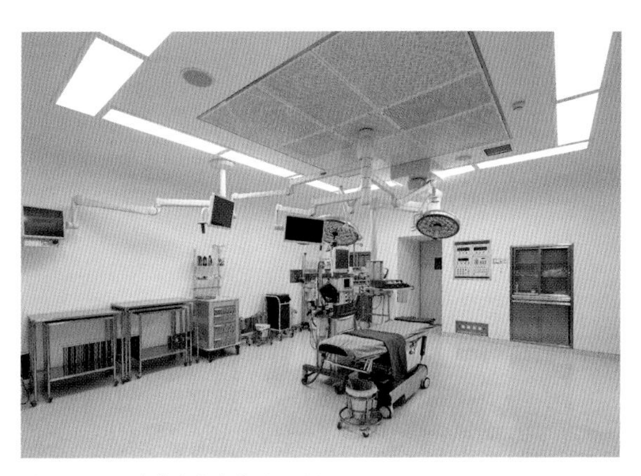

图 8-10 7# 楼手术室实景（组图）

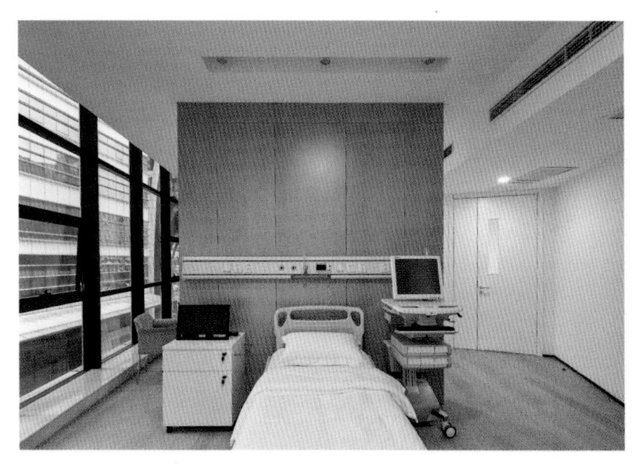
图 8-11 病房实景

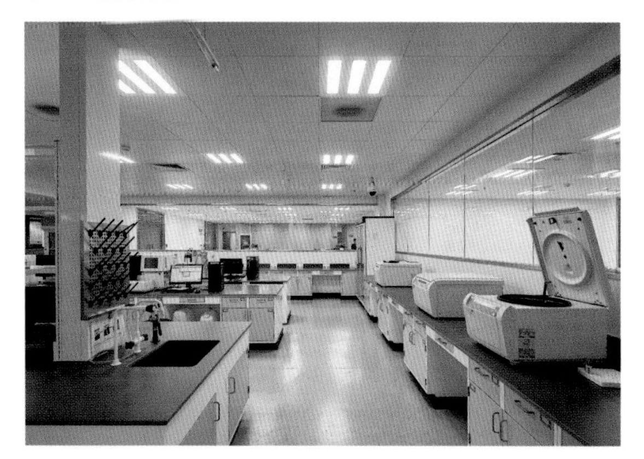
图 8-12 病理实验室

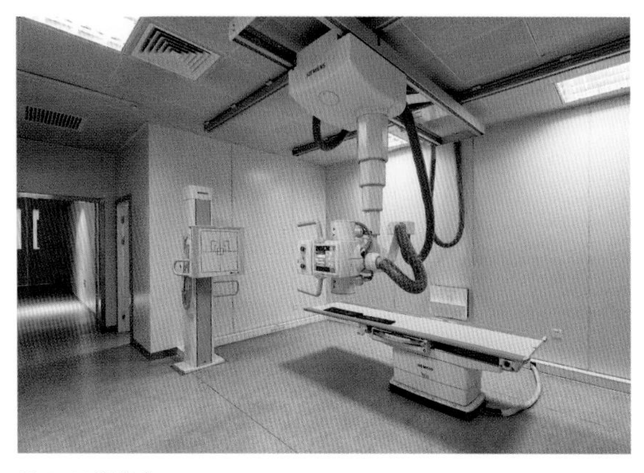
图 8-13 影像中心

圾；③结构拆除按自上而下、先水平后竖向的原则进行，防止因拆除处理不当对结构造成不利影响。

（四）移动医疗的应用

随着互联网技术发展，信息技术与医疗技术的需求相结合，移动医疗应用越来越丰富，如移动查房、移动护理、院内患者跟踪监护、患者自助应用、移动挂号及支付、远程移动会诊等。

①移动查房。PDA、平板电脑、移动查房车替代传统医生查房使用的纸质病历夹，患者健康档案能更加及时、完整、快捷地进行更新，避免差错，提高工作效率和质量。

②移动护理。移动护理的手持终端PDA轻便小巧，病区护士可以随身携带，在护理工作中，通过PDA扫描患者腕带、床头卡进行身份识别，准确无误地开展护理采集和治疗工作，减少差错。

③预约挂号。开通官方微信和APP手机挂号方式。患者通过手机挂号后，医生工作站即可显示该患者信息，患者可以按照诊室的叫号顺序就诊，无须到挂号大厅的服务窗排队取号，从而大大缓解医院门诊大厅的挂号压力，缩短患者在医院的停留时间。

④院内患者跟踪监护。建立院内患者跟踪监护管理系统，能有效节约医院人力资源，通过系统自动报警功能防止特殊患者走出安全区域，另外可以便于患者求助，准确定位，避免因寻找患者耽误救治时间。

⑤患者自助应用。患者可以进行点餐服务、上网服务、住院信息查询、吃药就医提醒等服务。

⑥远程会诊。带有医用PDA功能的手机，可以随时随地查看医学检验图片等信息，实现远程移动会诊。

（五）医疗标识、导向系统的设计

根据建筑和室内设计风格，采用与之统一的标志设计，造型简洁，内容清晰易懂，这不仅能提升医院形象，也能减少导诊等人力资源的浪费。按照由外到内、由大到小、由先到后的顺序而设计。

一级导向：医院大楼、大门、院名标识、道路指引和分流表识、医院楼宇标识、户外形象标识。二级导向：医院楼层索引及平面图、大厅及走廊标识、医院公共服务设施等。三级导向：各医院科室、护理单元、各行政后勤部门；四级导向：各房间门牌、各服务窗口、公共服务设施等。

五、未来改造性高端医院的趋势

①医疗观念变革。其他众多领域变化对医疗行业或多或少产生影响，主要体现在医疗观念、疾病筛查及治疗的手段等方面，而这些变化必将对医疗建筑设计提出更新要求。在美国，医疗观念出现以预防为主的趋势，预防远比治疗重要。便携式移动设备可以对人体的运动量，身体各种指标进行检测，并提供初步的分析结果供使用者参考。类似各种新手段、新观念的不断出现，对医疗建筑设计的影响是显而易见的。

②病房趋向单人间。病房单人间可增加私密性，使病人免受打扰。单人病房模式下，患者家属可以随时探病、陪护，护士台通过计算机平台对病房患者进行监护。

③小型化和专业化。随着移动医疗、远程医疗的应用，新型医院模式可能更集中在护理、新生儿接生、新生儿护理、大型手术和重大传染性疾病治疗等领域，常规的精神科、病理科、儿科、急症护理等科室主要设置社区，而不是集中在一个大型医院上，小型化和针对性越来越明显。社区卫生服务组织、综合医院和专科医院合理分工的医疗服务体系将逐步形成。公立医院做综合、私立医院做专科，公立医院保基本、私立医院做高端的趋势。

④全方位管家式服务。透过科学化的检测、分析、评估、预测、干预、追踪等步骤，形成“定制化”个人健康管理方案，以“全时态”的管家理念，透过直营特约体检中心智能化健康管理系统，先进高端的检测仪器，国际级的专家阵容，提供客户“一站式”“全方位”的健康医疗服务体验。以酒店会所式的空间规划、以客为尊的服务理念，从健康检查项目的安排制定、健康及心理状况的评估、运动指导、营养指导、中医体质的调理到抗衰老的预防，至国际专家会诊、医疗转介等，提供完整的、全方位的健康管理服务。

⑤重视后期的高效运行。医院建设不仅要考虑医院建设成什么样，还要考虑到建成之后如何运行，甚至要考虑到未来二三十年的发展问题。关于医院的高效运行，需要一支成熟的运行团队。

六、结语

对于北京优联健康医院的改造，我们通过对现场办公布局的仔细分析、合理规划，使其符合医院流程、布局和规划要求。高流量科室设置于底层，高私密性设置于高层，相关科室设置于同层或者相邻楼座，可灵活调节功能的设置在同层，通过对各层流线设计和功能优化布局，合理利用既有建筑垂直交通流线，使其满足医院要求，达到高效利用的目的，为广大患者提供更优质的医疗服务。

项目信息及版权所有

北京优联医院改造设计
业主：北京慈航投资基金管理有限公司
建设地点：北京市朝阳区
建筑设计：北京建院约翰马丁国际建筑设计有限公司
概念设计：香港利安设计集团
室内设计：北京建院约翰马丁国际建筑设计有限公司+北京建院装饰有限公司
设计指导：韩德民(中国工程院院士、耳鼻咽喉头颈外科学领军人物)
设计团队：1. 周彰青 2.蔡思 3.朱颖 4.杨超 5.何小萌 6.杨敬华 7.王鹏 8.张涛 9.邢向东 10.吴芳菲 11.付晓琳 12.韩起勋 13.郭小燕 14.张建 15.刘昕 16.杨旭 17.赵欣然 18.李轩 19.张金玉 20.杨一萍 21.朱义秀
基地面积：80000平方米
总建筑面积：69000平方米
项目状态:已建成
设计时间：2014—2017年
建成时间: 2017年
摄影/图片版权：北京建院约翰马丁国际建筑设计有限公司
撰文：周彰青 蔡思 韩起勋 刘昕 杨一萍

玖记 Note Nine

建筑有机更新
Organic Renewall of Architecture

青岛体育场、柳州银泰中心更新改造设计
Renovation Design of Qingdao Stadium and Liuzhou Yintai Center

创意理念一方面源于城市自身文化特征的保留，另一方面源自城市产业的再创新。
找到城市发展的新视角，
从而将城市文化策略在文化规划层面上扎实展开。
不仅有城市专属文化区，还要有展示平台，
这是新旧城市、大小城市复兴之关键。
所以，本记的理论与实践探索，
旨在厘清文化空间的类别并找准文化空间资源优化配置的要点。

Creative ideas come from the cultural heritages of cities on the one hand, and also originate from the constant innovation in urban industries on the other hand. These ideas help cities seek for development from new perspectives. As a result, urban cultural strategies can find their concrete expressions in cultural plans. Cities need not only the exclusive cultural zones but also the presentation platforms, which are pivotal for the revival of cities, old or new, small or large. Therefore, this paper makes theoretical and practical explorations to classify a variety of urban spaces and optimize the allocation of cultural space resources with clear-cut priorities.

青岛体育中心主体育场改造设计

建筑作为城市的基本组成单元，如同细胞之于动物体的关系一样，都有新陈代谢和有机更替的环节。对于建筑来说，这一环节不但与建筑自身的使用寿命相关，更是与社会的经济发展状况和城市需求增长密切联系的。随着中国城市建设的飞速发展和人民生活水平的日益提高，城市建筑尤其是部分公共建筑的设计理念和功能使用已经无法满足现代化生活的更高要求，而且这部分公共建筑已经在城市构架中形成了可识别的节点，并在人们心理范畴中成为城市生活的一部分，如采用拆除重建的方式，势必将该建筑及其所在区域的历史和时间特性消除，形成城市文化中的断章。因此对这类建筑，更新改造成为了更为适合的处理方式，且经过更新改造后的建筑不但功能上焕发新的生机活力，同时也延续了城市的历史文脉和公众记忆，并使得该建筑具有成为城市标志性建筑的可能。这一点，在诸多建筑改造中都得到了很好的体现。

青岛体育中心主体育场改造设计就是基于上述观点的一个鲜明案例。

青岛体育中心体育场位于青岛市崂山区银川东路3号（青岛体育中心园区内），始建于1996年，1999年建成，占地面积约4.5万平方米，座椅约4.9万个，曾作为足球联赛场地使用。由于年久失修、结构老化和海风腐蚀，各专业及公共体育设施破损严重，无法使用：观众席屋顶膜结构已达使用寿命急需更换，膜结构的钢支撑存在锈蚀情况，各类功能流线交叉不能满足赛事功能要求和消防疏散要求，体育场现状无法

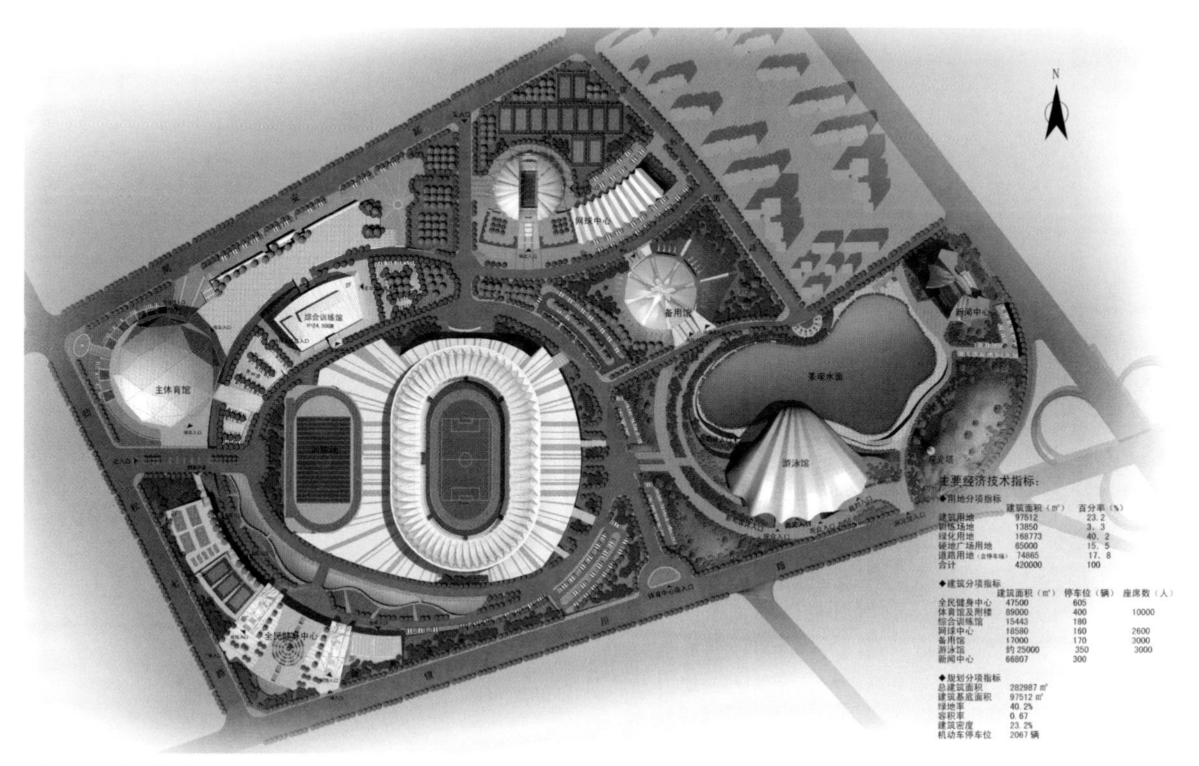

◆用地分项指标	建筑面积（㎡）	百分率（%）
建筑用地	97512	23.2
训练场地	13850	3.3
绿化用地	168773	40.2
硬地广场用地	65000	15.5
道路用地（含停车场）	74865	17.8
合计	420000	100

◆建筑分项指标	建筑面积（㎡）	停车位（辆）	座席数（人）
全民健身中心	47500	605	
体育馆及附楼	89000	400	10000
综合训练馆	15443	180	
网球中心	18580	160	2600
备用馆	17000	170	3000
游泳馆	约25000	350	3000
新闻中心	66807	300	

◆规划分项指标	
总建筑面积	282987 ㎡
建筑基底面积	97512 ㎡
绿地率	40.2%
容积率	0.67
建筑密度	23.2%
机动车停车位	2067 辆

图 9-1 园区规划总图

图 9-2 东立面效果图初始方案（上）及实施方案（下）

满足现行功能标准、安全规范的要求。原项目已经完全不能使用。

针对项目是采用拆除重建，还是维修改造，政府主管部门、业主、设计团队进行了多次的讨论，如拆除，现有场地条件依据最新的体育设计规范无法在原址上重新建设同等规模的体育场，只能移至其他区域重建。而现状体育场虽已无法使用，但其巨大的规模和独特的形象已经成为青岛崂山区乃至青岛市的标志性建筑物，并在精神层面承载着周边人民生活和中国足球的某些特定记忆，拆除势必也将引起社会舆论的不同声音。综合以上各种因素，同时考虑到体育建筑

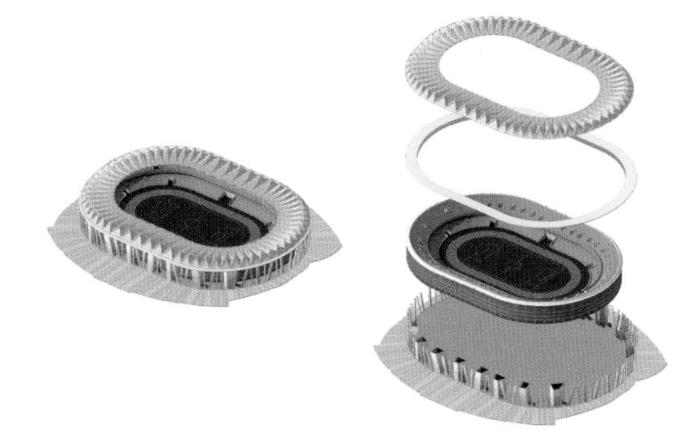

图 9-3 立面构成分析

自身就存在依据赛事进行局部改造的特殊性，项目最后采用了维修改造的方式。

在马国馨院士的指导下，设计团队对体育场进行了全方位的改造设计。在体育场改造定位方面，马国馨院士结合体育场现状并针对国内缺乏足球专业场地的现实情况，提出将体育场改造成为专业足球场，颇为遗憾的是，这一建议因为诸多因素而未被采用。最后体育场定位为综合性的体育场。其改造的主要内容包括加建疏散平台，调整内部流线，增加疏散出口，更换屋顶张拉膜结构膜材，对钢结构进行维护，部分结构拆除并进行加固，重建主席台、媒体看台、评论员席，整修装饰观众看台，更换全部座椅，重新划分功能布局，重新装修所有房间，改造外立面，改造体育场的场地、照明、扩声、记分牌等，改造给排水、消防、强电、弱电、暖通系统，同时改造室外道路、广场、停车场、绿化、照明及管网等工程，并购置草坪养护机械、体育器械、办公家具等。

而这些改造措施中最重要的两个部分就是内部功能的重新划分和外立面改造。

原体育场东西南北项目4个看台由于历史原因是独立建设的，互不联系。南北看台下方为酒店，西看台观众流线和运动员、工作人员、媒体流线互相交叉。为了使体育场功能达到现代体育赛事的要求，设计方进行了大胆的功能改造。

首先，通过对内部功能房间的重新整合，在体育场的上层看台和下层看台的入口处，即体育场的2层和4层形成了环形的入场廊道，体育场各个方向的看台通过这两个环形廊道完全连接起来，充分满足了观众入场散场流线和消防疏散的需求。

其次，将原来与下层看台流线混合的贵宾席和包厢层独立开来，在3层形成了贵宾和包厢的独立区域并设置独立的入口，使体育场内的看台座位的层级划分清晰明确，适于运营和管理。

最后在体育场西侧加建了疏散平台，疏散平台上

图 9-4 实施方案西北侧效果图

方为观众的入口区域，疏散平台下方为运动员、工作人员和媒体的入口区域，这一划分使体育场在体育功能上充分满足现代体育赛事的需求。而原来东侧疏散平台下方的室内停车场依据园区内的总体规划改造成了体育服务的配套设施，弥补了园区内配套设施的不足，同时为场馆运营提供了良好的条件。

通过上述的功能处理，原本功能复杂、流线混乱交叉的体育场在功能上变得清晰而有秩序。这也充分表明了设计团队对这种涉及大规模人员疏散的复杂公共建筑的成熟处理手法。

外立面改造是体育场另外一个重要的部分，在设计初期，设计团队在马国馨院士指导下提交了多版外立面改造的方案，其中较为理想的一版立面方案对体育场立面进行了全新的设计和包装，新颖实用而又不失体育建筑的气魄和力度，既解决了长久以来体育场屋面排水的问题，又对体育场的各入口区域进行了空间上的划分和限定。这版方案由于造价和实施难度的问题未被采用。最后设计团队只对现有的体育场外立

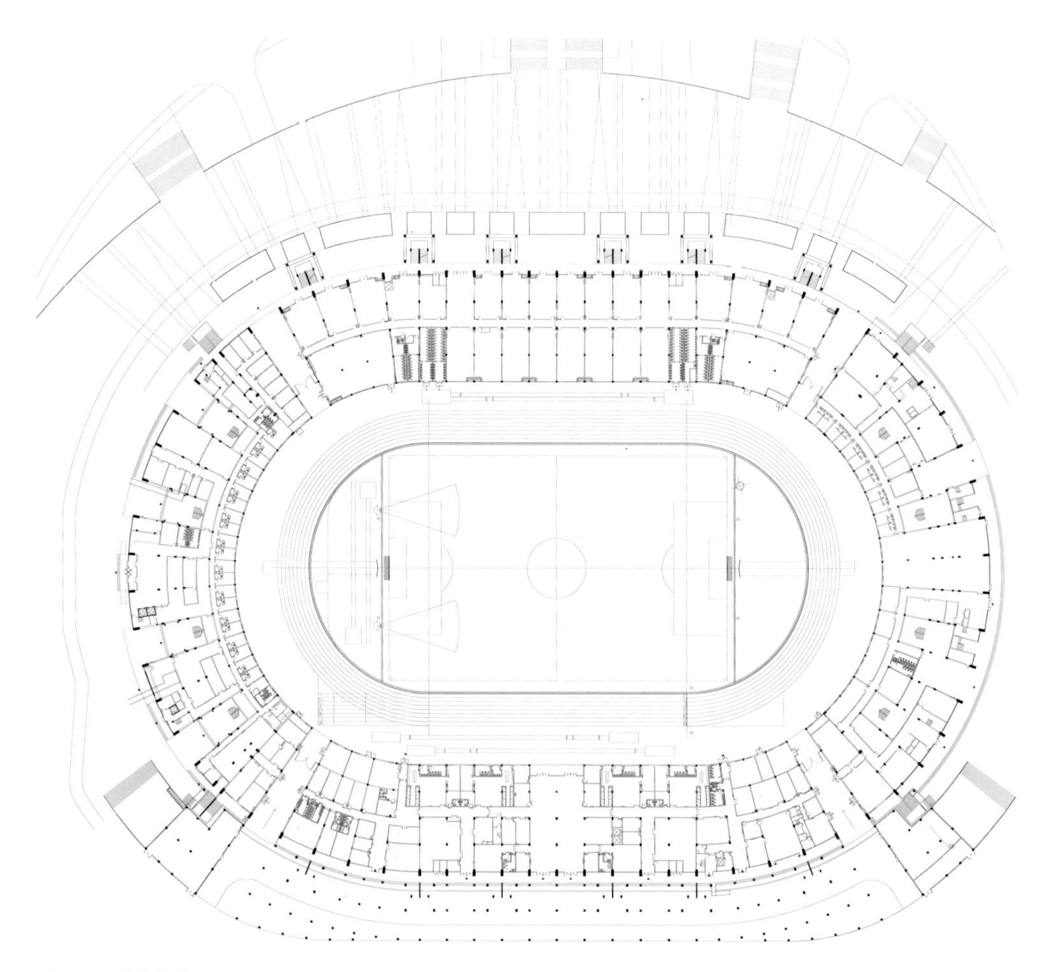

图 9-5 一层平面图

面进行局部的处理，把原有略显琐碎的室外疏散楼梯进行了实体化的处理。这样改造后的体育场外立面最大程度保留了原有体育场的立面特点，并增强了体育建筑的厚重和力量感。这样的处理，表现了设计团队对城市建筑更新的态度，既最大限度的尊重和保留建筑的历史记忆，并通过成熟的手法对其弥补和提升，从而使城市建筑的时间感和文脉得以传承和延续。

经过精心改造并对公众开放的新体育场可容纳40000人，为甲类体育建筑，具备体育场、运动员接待中心、运动员驻训用房、办公、餐饮等功能。其具备承办国际足球单项赛事、国内足球最高赛事、国内田径一类赛事的能力，同时也承担着开展青岛市重大节庆活动的重任。在项目的改造过程中也有一些遗憾的地方，如：在体育场的室内设计环节中，设计团队在体育场所有半室外的环廊顶部设计了格栅式的吊顶，并将体育场内指示标识重新设计与墙体和地面融为一体，形成整体式的装饰。但由于预算、施工工期和施工技术等问题均未能实现；体育场存在已久的视线问题，设计团队创意性地提出将体育场场心垫高的根本解决方案，也因预算问题无法实现。在体育场今后的使用过程中，伴随着新的赛事需求，上述遗憾之处或许能得到弥补。

建筑的更新改造从某种程度来说比建造新建筑要复杂和困难，因为它不仅仅需要此时此地与环境与功

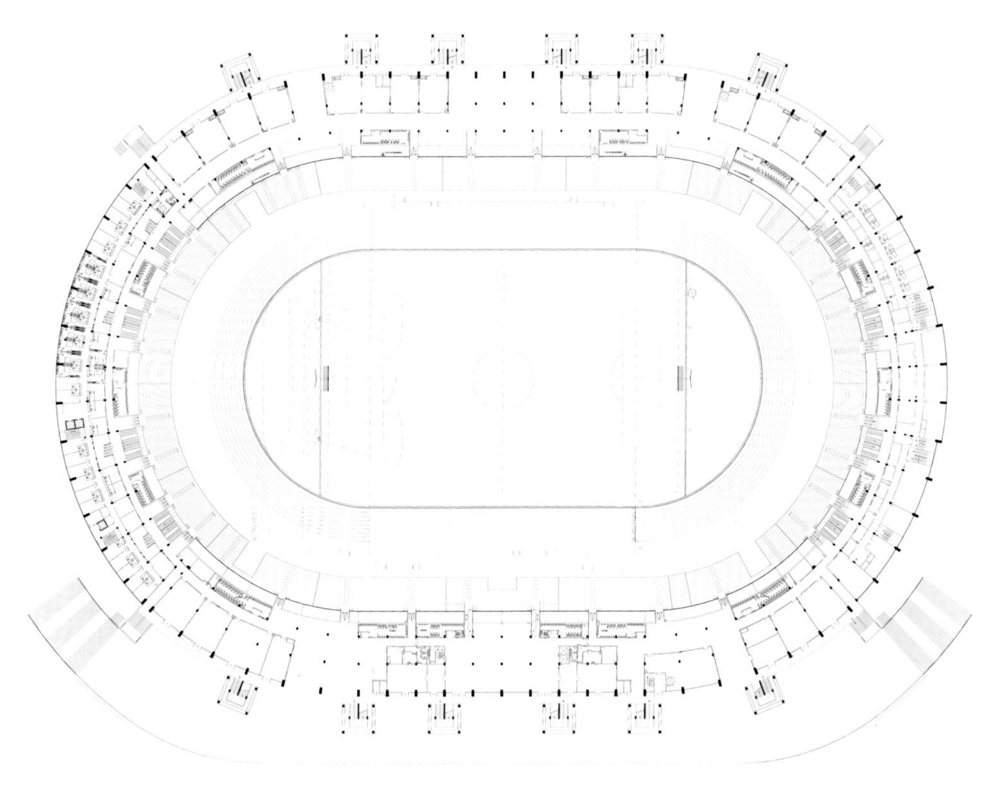

图 9-6 二层平面图

能的契合，更需要从时间的维度从文化和心理的范畴去考虑问题，建筑初始的记忆和重生的状态是彼此渗透和交融的。或许那种历久弥新的沧桑感才是建筑真正感人的部分。

柳州银泰中心改扩建

"适者生存"是达尔文自然选择学说的一个组成部分，意为自然选择使那些较不适应环境的个体淘汰，最能适应环境的个体得以保存和繁荣。1864年英国社会学家斯宾塞在达尔文的《物种起源》出版后不久，根据达尔文的生存竞争的思想，提出适者生存的概念，以描述自然选择的原理。同时进一步将这个生物学的概念引入社会历史领域，认为社会进化过程如同生物进化过程一样，生存竞争的原则起着支配作用，适者生存同样有效。而纵观中外建筑发展史、城市化进程，适者生存同样是建筑发展、城市更新繁荣必须遵循的不二法则。

达尔文的生物进化论认为在生存斗争中，具有有利变异的个体，容易在生存斗争中获胜而生存下去。反之，具有不利变异的个体，则容易在生存斗争中失败而死亡。这就是说，凡是生存下来的生物都是适应环境的，而被淘汰的生物都是对环境不适应的，这就是适者生存。达尔文把在生存斗争中，适者生存、不适者被淘汰的过程叫作自然选择。达尔文认为，自然选择过程是一个长期的、缓慢的、连续的过程。由于生存斗争不断地进行，因而自然选择也是不断地进行，通过一代代的生存环境的选择作用，物种变异被定向地向着一个方向积累，于是性状逐渐和原来的祖先不同了，这样，新的物种就形成了。由于生物所在的环境是多种多样的，因此，生物适应环境的方式也是多种多样的，所以，经过自然选择也就形成了生物界的多样性。

细观世界各地区城市和建筑发展史，也同样遵循了自然选择的发展规律，形成了现在各具特色的城市风貌。因此，以"适者生存"的理念为指导，研究既有建筑改建中各种错综复杂问题，寻求正确解决方法加以实施，将是符合自然规律的行之有效的法则。

先找出不"适"的问题，提出解决方法，以使其具有更强的生存能力。间接延长建筑寿命，提升城市活力。

柳州银泰中心改扩建就遵循了这一适者生存的原则。

图 9-7 从柳江看现状楼　　图 9-8 现状内中庭　　图 9-9 现状西入口

图 9-10 西侧效果图

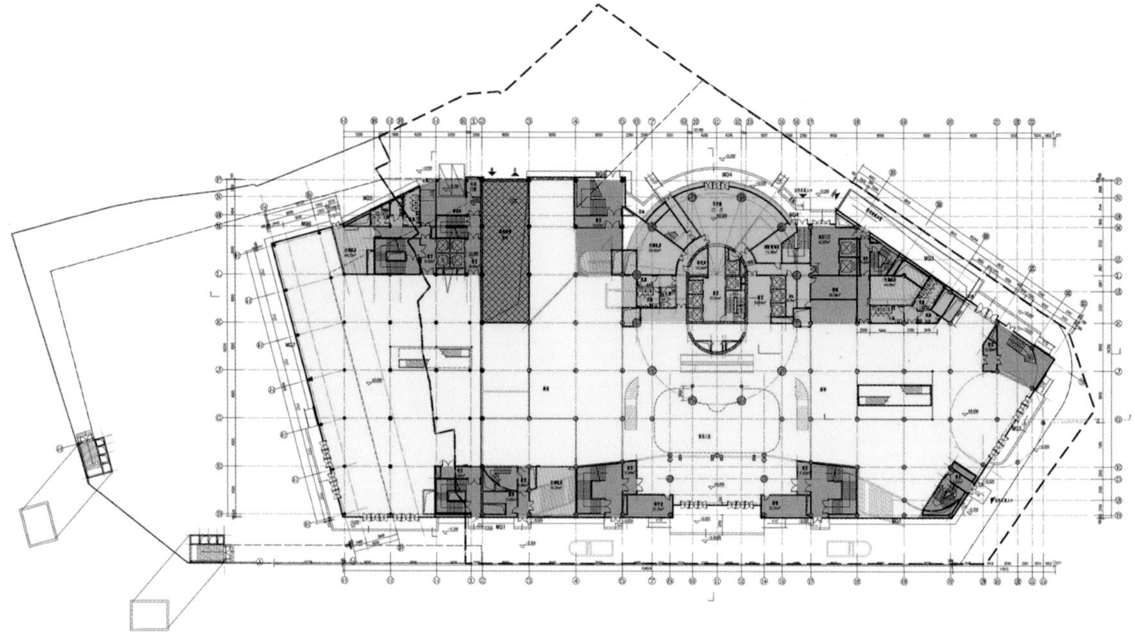

图 9-11 一层平面实施方案

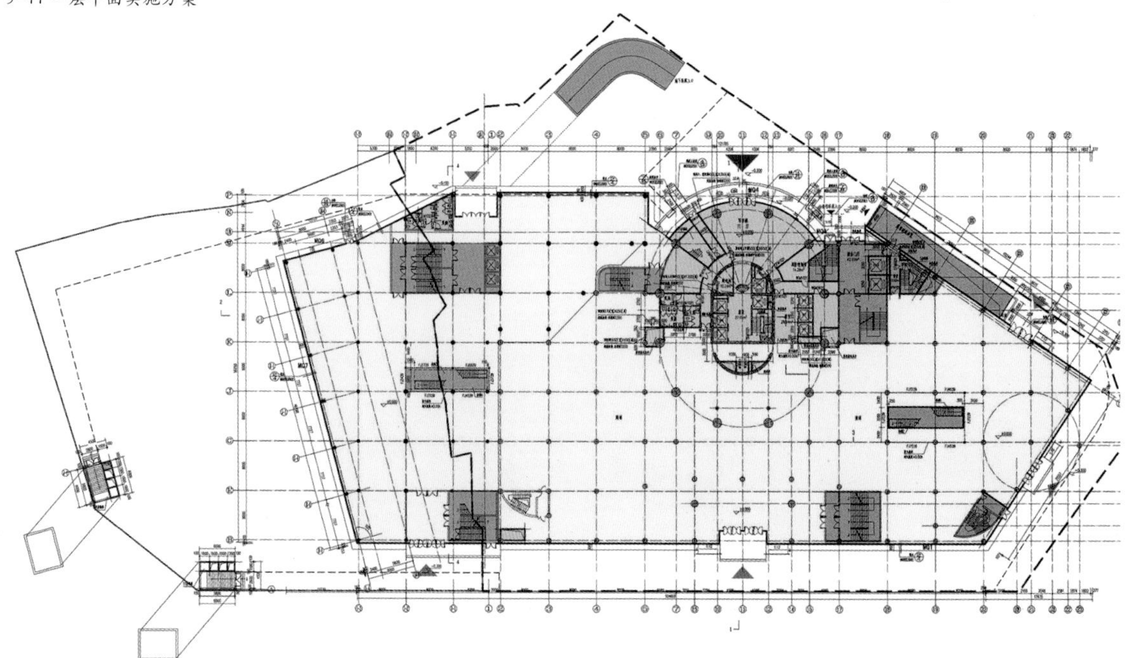

图 9-12 一层平面原方案

（一）缘起

2012年我们接到银泰委托进行柳州银泰中心的深化施工图设计，查看资料发现当时距该项目启动已20年了，距整体结构施工完成14年了。也就是说这项目已建造了20年还没竣工，号称“广西第一烂尾楼”。想来其中必有缘由。我们首先对现状情况和图纸资料进行分析。

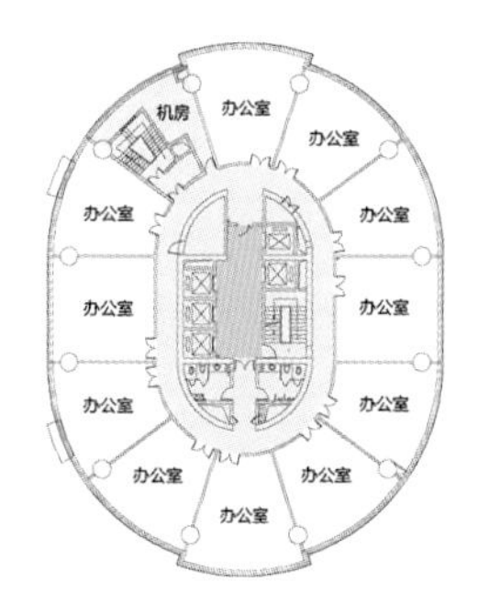

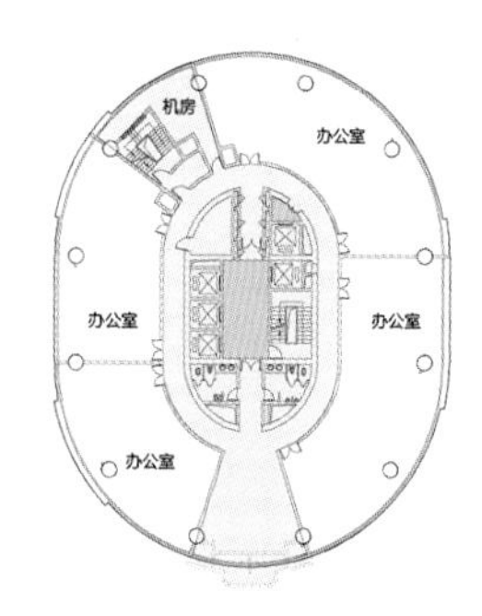

图 9-13 二层平面实施方案（左）及原方案（右）

（二）项目现状

本项目位于柳州市中心南北主轴线上，北面向柳江，为正在实施的“柳江百里画廊”重要景观节点，是柳州市中心区柳江南岸城市轮廓线的至高点，对构建新时期柳江市城市新风貌、体现城市特色，影响巨大。

本项目西临鱼峰路（道路红线宽40米），北临驾鹤路（道路红线宽32米），东临规划路（规划路宽6米），东南临太平西街（道路红线宽12米），总用地面积11600.09平方米，其中，城市公共绿地1781.92平方米，建设用地9818.17平方米。

该楼当时定位是柳州重点商贸中心，功能复杂。总建筑面积约65734平方米,地下3层，地上33层超高层建筑，地下2~3层为车库，地下1层至地上7层裙房为商业、餐饮各种文体娱乐。塔楼为办公建筑，开工前改为四星级酒店。酒店大堂设在9层，顶层两层为观景餐厅。

1992年方案中标，1994年开工，因处理地基至工期延误，1995年施工到地面以上主体结构，1998年进行过主体结构竣工验收，同年在设备安装及外幕墙未完成，内装修未进行的情况下，因投资中断而停工。当时主体结构、二次结构全部完成，大部分安装及外立面工程已基本完成，此后在未维护状态下经历了多年风风雨雨。加上地处柳州市区繁华地段，不知不觉成为“广西第一烂尾楼”。

新加坡海峡置业公司此前购买银都大厦，将其更名为“新银都大厦”，并成立柳州新银都房地产开发有限公司，成为其新主人。2009年柳州新银都拍下项目北侧地块，作为新银都大厦的北侧裙楼扩建用地。银泰集团控股新银都大厦后，将在此建设银泰城项目。银泰城将被打造为规模化、时尚化、品质化的标志性商业建筑体，涵盖零售、餐饮、娱乐、办公、休闲等诸多功能。本建筑地上33层，地下3层。地下2层、地下3层为停车库，地下1层至7层主要功能为商业、餐饮、娱乐，塔楼为写字楼。

2011年该项目由当地设计院进行了改造、扩建施工图设计，出了报证施工图。报证施工图已经通过政府各部门审批，取得规划建设许可证。改扩建设计之后，原建筑东北侧外接地下2层、地上8层裙房的新建建筑，整体建筑面积为91099平方米。

（三）设计过程

2012年银泰集团接管此项目，委托我方设计团队进行以下深化设计。

该项目施工图设计时，平面业态布局未定，外立面效果也未达到银泰要求，结构加固改造缺深化设计，机电系统设计针对性不足。因此，要求我们本阶段施工图设计主要是结合银泰集团百货营销的具体需求，招商落位平面，在原施工图的基础上再作深化设计。

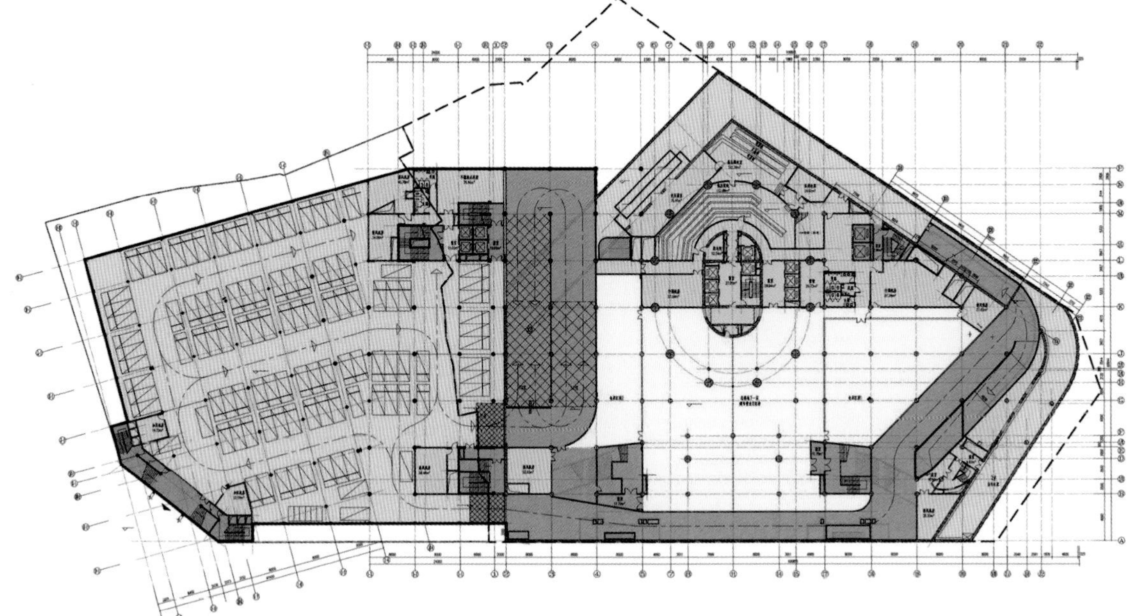

图 9-14 地下一层平面实施方案

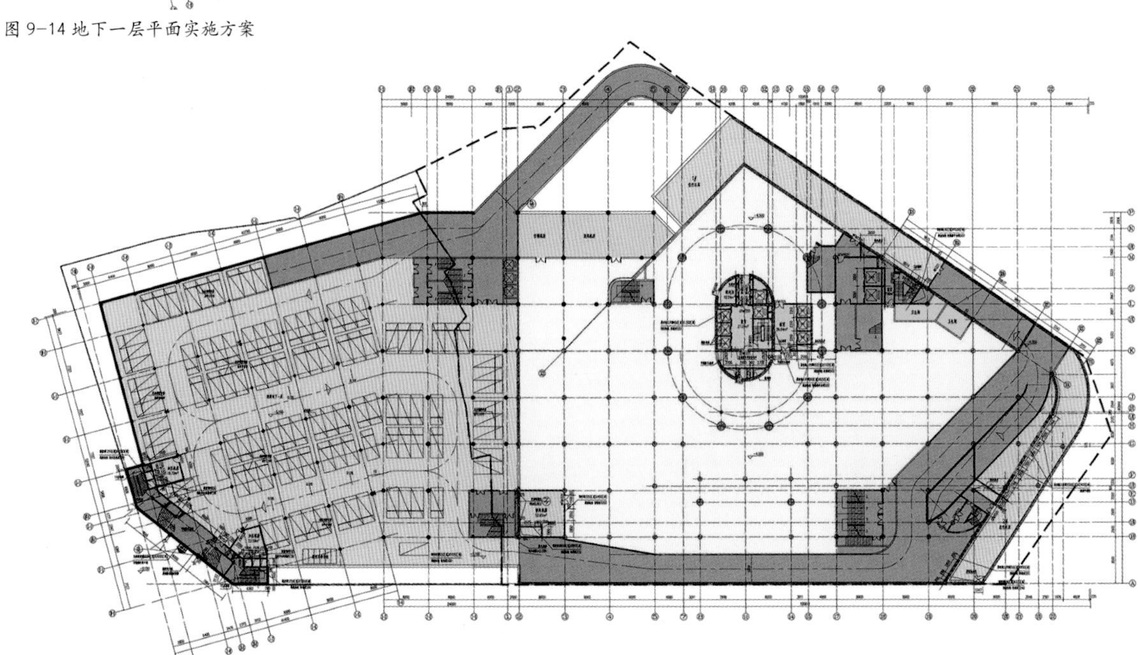

图 9-15 地下一层平面原方案

（四）发现问题

①图纸、资料不全：由于项目设计多次更改，未进行图纸资料归档，图纸资料混乱不全，且多处与现状对不上。如果只是简单在现有图纸上深化设计，会导致施工无法进行等问题。

②提供施工图与最新的使用需求矛盾：现状和改扩建图存在很多方案性的硬伤问题，若不改善会为这幢超高层地标性建筑带来永久性遗憾，甚至会影响其未来的发展方向，导致又一次烂尾。主要问题如下。

ⓐ汽车坡道位置问题。

现设于老楼北侧的8米宽双车道打断了东北部新建部分与原有部分地下层及首层的联系，造成地下停车不便、不经济，不利于首层商业内街的通畅。同时还在总图交通设计上对办公人流造成交叉，带来安全隐患。

ⓑ首层平面商业价值未充分挖掘问题。

首层沿街商业价值最高的位置设置大量楼梯、机房，尤其是首层沿鱼峰路和太平西街这些主要的商业

图 9-16 东南侧鸟瞰效果图

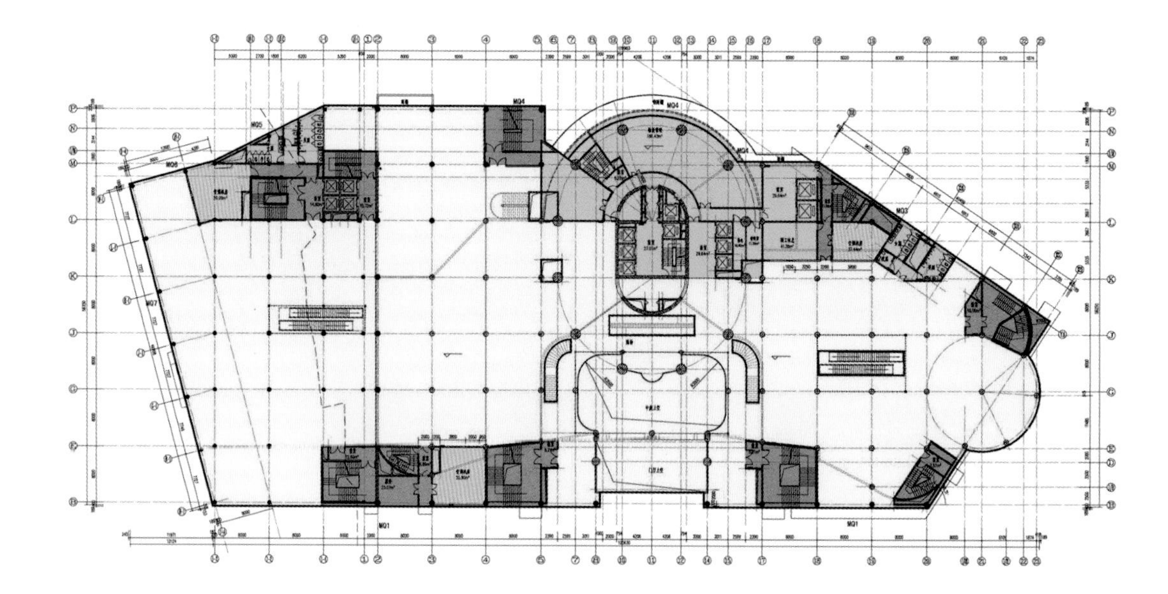

图 9-17 二层平面实施方案

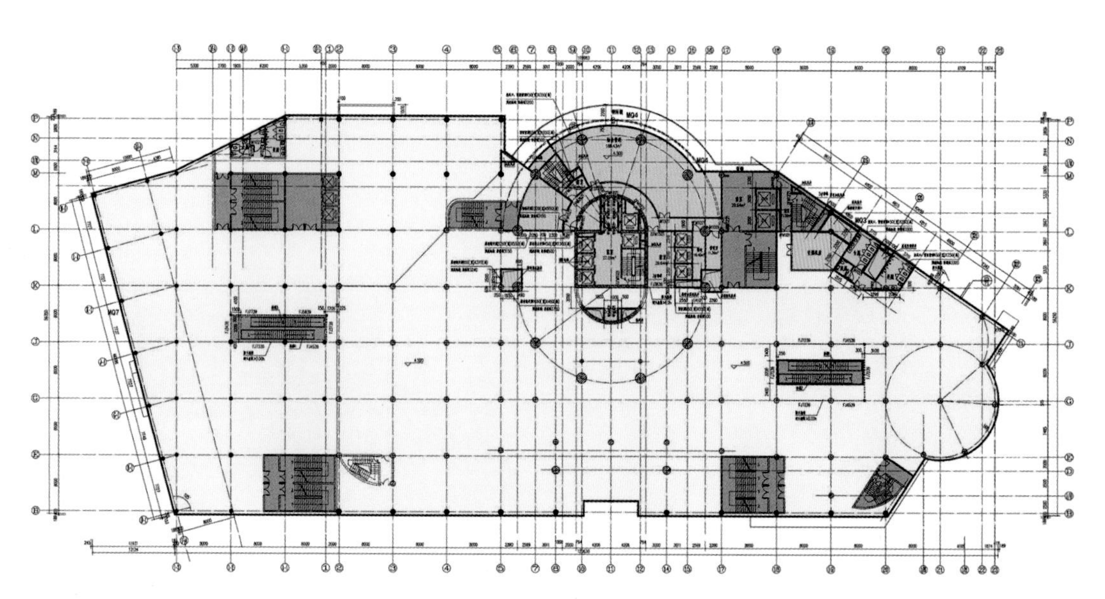

图 9-18 二层平面原方案

街处，破坏了整条街的店铺鳞次栉比的商业氛围，不利于未来吸引客流，降低了项目的经济效益。且楼梯设计过于随意，占用大量平面空间，完全可以精简整合。设备机房过多占用首层高商业价值平面面积，导致整条街只有两个入口。在商场关门时间此条街就没有吸引力、少了人气、失去活力，和当地老商业街比肩店铺，24小时都有经营的不夜街区，让人们流连忘返的、永远人流涌动的热闹喧嚣氛围相去甚远。同时也会造成客流的缺失，不利于未来经营。

ⓒ老楼中庭问题。

老楼1~6入口设6层挑高空间，原设计可能是希望有高大空间的大厅，18米宽、6~8米进深的狭小开洞，配25米多高的空间，完全不能达到目的，只会加大空间的局促感。同时将东北侧新加建的商业与南侧老楼裙房的商业联系打断，不利于各层内商业街的连通，无法形成合理的商业动线。

ⓓ新旧区新加自动扶梯位置问题。

原施工图在南北两处新加自动扶梯位置设置过于偏两端，导致尽端无法形成大型商业活动空间，使得整个商业层使用进深过于匀质，缺少变化，不利引入大型活动业态，限制未来动态的使用变化需求。

ⓔ高层塔楼电梯问题。

塔楼部分每层面积为959平方米，作为高度141.5米33层的超高层建筑有些偏小，而平面竟设置了三套核心筒，中心一套由四部电梯及电梯厅、一部楼梯、配电间、卫生间及一部货梯组成，且货梯厅为极为不合理的狭小的独立空间，另在建筑西侧主立面外墙处设两部观光电梯，此观光梯占用大量使用面积，且观光方向朝西侧商业楼，景观并不好。

由于核心筒电梯已安装，新加33层观光层没有电梯直达，而是又在西侧破楼板设步行梯上到33层，既占用使用面积，又达不到无障碍和舒适。

此阶段，银泰只是委托我们按目前图纸根据商业招商图落位到原施工图中，使机电专业能深化设计，为下一步的装修设计提供条件。但看到带有如此多硬伤的图纸，我们如鲠在喉，职业道德让我们决定提出一套调整方案给甲方。

（五）改造调整方案

①设计原则：在原有已审批的建筑轮廓范围内设计，满足政府要求。

②重新完善现状图纸：各专业工程师深入现场，参照业主提供图纸结合现状，重新绘制现状电子图，以保证设计依据的准确性，此工作需要设计师大量的投入和细致的工作，为下一步工作打好基础。

③提出改造方案要点。

ⓐ移走车库坡道，一举多得。

在总图上优化中部汽车坡道位置，将其移到东南侧用地入口处，使车行来客直接在口部进入地下车库，从而做到人车分流，提高人行的安全；楼内车道的移出使地下3层停车空间加大，增加停车量，地下3层通过西侧新设直坡道顺畅到达新加地下2层；地下2层新旧区停车空间连通，通过中心直坡道连接，大大减少原有坡道占用的面积，增加了停车数；地下1层老楼区商业面积加大，同时可直接连通新加地下1层车库，方便使用；减少坡道占用首层面积，保障新旧商业空间使用的完整性，使得新旧建筑在内部浑然一体。

ⓑ首层的策略，商业价值最大化。

由于首层在商业用房中是价值最大的，因此要尽量减少不必要的配套用房。将不影响使用的机电用房布置在相对商业价值偏低的其他层，从而释放出更多商业面积。

ⓒ好好设计楼梯，腾出的不只是面积，是商业街的热闹繁华。

商业建筑由于人流量大，楼梯的疏散量自然可观。且首层楼梯位置以能直接通往室外为最优，而过多的楼梯沿街布置，必然影响首层商业与外部的联系。

通过研究楼层高度、设计规范要求，开动脑筋，发挥创新性思维，提出将原设计的三跑楼梯改为两跑到四跑楼梯，使得楼梯面积大大减少，使商业面积最大化、沿街商业界面最大化提供了条件，有利于开设沿街小店铺，形成原老街店铺比肩相邻的热闹氛围。

ⓓ封上老楼中庭，连通新旧商业空间。

将老楼2~6层高中庭开洞补上楼板，使新旧楼层成为一个通畅的整体，一个开放式空间，为未来的使用提供多种可能。

ⓔ合理布置自动扶梯位置，是项目成败的关键因素，同时可提供层层通达的购物体验。

此项目商业自地下1层至地上7层，扶梯设置通达各层；首层靠近口部，便于人流进入其他层；两部扶梯均布置在老楼区同一轴跨间，考虑北端新建区进深都在20多米，适合预留出相对大些的店铺与老楼区的内街式小店铺形成变化，同时能提供使用的灵活性。

ⓕ 顶层提供高大空间，为多样使用提供可能

顶层由于没有了上部楼层的制约，在不需要过多增加造价情况下，尽可能在设计中采用局部大跨度（适当位置减少柱子），增加空间自由度，使得项目在6~7层设置多厅影院成为可能，也为未来的应对市场变化，持续发展提供条件。

ⓖ塔楼电梯两消一转：小的修改致大的改善，使得这建筑的潜力充分发挥。

取消西侧两部观光电梯井，该电梯厅改为办公室，增加了办公面积，提高了每层的使用率，让办公尽享柳江全景；同时巧妙地将货梯门洞朝向逆时针回转90度，与大楼电梯厅共用前室，将走道两侧管井移到原货梯三角厅，形成宽大通畅电梯厅空间，并与原有4部客梯组成5部电梯满足超高层电梯数量要求；此货梯借调整机会，直接通到33层，解决了原该层电梯无法到达之缺憾；取消西部观光梯弱化了西立面的布局，为靠近柳江的北立面成为主体立面提供了条件。

（六）调整方案

我们在新方案基础上又对整个楼的消防设计、节能设计、绿色环保设计按目前国家规范重新进行梳理，结合业主提供的外立面方案进行施工图设计，以使本项目符合时代的要求。

建筑物内部功能重新规划，迎合了当代人的需求，其内部的各种配套设施优化设计和整改，使建筑使用空间得到了充分有效的利用。

外立面以时代感鲜明的新颖形象，体现当代柳州人的精神气质和文化品位，同时又与“历史文化名城”柳州的传统文化有内在的关联。由全玻璃幕墙塑造的、抽象雕塑般的简洁建筑形体，似船舶、似花苞、似百里柳江山水间晶莹剔透的宝石——可给人多种联想和美的感受。与对岸“柳州风情港”隔江“对歌”，与山水环境和原有城市建筑群融合，成为鲜明的地标。

适者生存理念的设计导向、可持续发展的目标贯穿于大楼的方方面面。开放式的建筑设计，延长了建筑寿命，全面提升了建筑性能，改善了建筑品质，改变了建筑属性。随着建筑内部及外部一系列精心的规划设计，这栋搁置许久的标志性建筑物获得了重生，名副其实地成为影响新时期柳江市新风貌的构建和体现柳州市城市特色的地标式建筑物。

（七）对更新改造的思考

①更新改造的必然性。

既有建筑量大、能耗高现状：据前瞻产业研究院发布的《2013—2017年中国智能建筑行业市场前景与投资战略规划分析报告》中的数据显示，我国目前既有建筑面积超过500亿平方米，90%以上是高耗能建筑，城镇节能建筑占既有建筑面积的比例仅为23.1%。权威机构调研说明，如果达到同样的室内舒适度，中国单位建筑面积能耗是同等气候条件发达国家的2至3倍。无论是资源、环境的现实压力，还是群众对居住环境舒适度的迫切要求，既有建筑的节能改

造势在必行。

建筑品质低、弱、差现状：大部分既有建筑是改革开放之前或20世纪90年代之前建造，受当时经济和社会发展等因素设计标准低，防灾能力较弱， 使用功能差。

建筑使用内容变更的需要：随着时代的变迁，人们的生活发生着天翻地覆的变化，建筑的使用功能随之调整也成为必然。

②更新改造关注的问题。

如何来改造这样一大批既有建筑，延长使用寿命、提高使用功能和室内外环境是建筑行业要解决的一个很大的问题；如何改造既有建筑以满足现代生活的需要，赋予旧建筑以崭新的生命；如何判定定位决策的准确与否；建筑设计与室内装修、景观如何协调；如何恰当处理现存建筑；这都远非保存都市风景、历史遗迹建筑那么简单。如何利用新技术新材料满足改建中节能、环保、舒适、经济问题，达到旧建筑与新技术的完美结合；如何保留历史印记，利用现代技术，完善老旧建筑设施；如何综合利用，让新旧和谐；这些则是考验设计师综合素质的试金石。

既有建筑改建需要关注三方面：时间、地点、建筑。

时间：指不同时期的建造的建筑。具有不同的社会价值、文化价值、经济价值、生态价值等。地点：指不同地区的建筑，建筑与环境有不可割裂的联系。建筑：指既有建筑本身是何种结构体系。建筑师应因地制宜，实事求是，结合既有建筑具体情况制定改造方案。

既有建筑的改造还要避免同质化。现代科技的进步、市场经济的发展，已经将地球变成全球村。千城一面的城市，由混凝土和幕墙堆砌建的方盒建筑替代了自然生长的老旧城区，也割裂了与原有生活和历史的联系。

既有建筑改造提供了新建筑无法体现的东西，如历史文脉、富有意义的个性特征。精心地保存历史痕迹，挖掘建筑潜能，满足新的需求，以实用和审美、高品质和科技，精心再造，恰当处理，化腐朽为神奇，让旧有建筑获得重生。这是改造的重中之重，也是城市有机更新的重要原则。

善待历史，体会当下，面向未来，是我们建筑师在城市有机更新和建筑改造中应有的思维方式。

项目信息及版权所有

青岛体育场
业主：青岛市国信体育开发有限公司
建设地点：山东省青岛市
建筑设计：北京建院约翰马丁国际建筑设计有限公司
联合设计：北京市建筑设计研究院有限公司第二设计院
设计指导：马国馨（中国工程院院士、全国工程勘察设计大师）
设计团队：1.王鹏 2.朱颖 3.李诗云 4.周彰青 5.朱琳 6.邹雪红 7.葛亚萍 8.韩涛 9.曾劲 10.王皖兵 11.王皆欣 12.谷凯 13.赵伟 14.熊进华 15.姜建中 16.杨春红 17.刘国
基地面积：45000平方米
总建筑面积：84193平方米
项目状态:已建成
设计时间：2012年
建成时间:2014年
摄影/图片版权：北京建院约翰马丁国际建筑设计有限公司/北京市建筑设计研究院有限公司

柳州银泰中心设计
业主：柳州新银都房地产开发有限公司
建设地点：广西省柳州市
建筑设计：北京建院约翰马丁国际建筑设计有限公司
联合设计：和桥设计
设计团队：1.张彤梅 2.杨林 3.王晓光 4.王彪 5.胡益莎 6.罗超英 7.雷晓东 8.田玉香 9.刘春玲 10.时雅洁 11.孙宏雷 12.赵伟 13.尹鹏 14.杨春红 15.姜建中
基地面积：11600平方米
总建筑面积：91099平方米
项目状态:已建成
设计时间：2012年
建成时间:2014年
获奖情况：
摄影/图片版权：和桥设计/北京建院约翰马丁国际建筑设计有限公司
撰文：王鹏 张彤梅

2048
Vanke

拾记 Note Ten

世博会万科馆

Vanke Pavilion of 2010 Shanghai Expo

“麦垛”——秸秆板材料探索性实践

“Straw Stacks”- Exploratory Practice of Straw Board Material

尽管2010年上海世博会已经过去了7年，
但作为设计者我们没有忘记曾经为中国世博遗产做出的探索，
我们值得记忆下作为建筑师在这个项目中所领悟到的责任。
万科馆的主题是“全球变暖”，
基于对这一主题的认同，我们将其作为了建筑创作的出发点。
特别由建筑及建材生命周期CO_2排放量的研究结论，考虑到万科馆是临时场馆，
于是就大胆尝试用秸秆板“盖房子”——将其作为建筑结构材料，
然后以材料→结构→空间的推演为基础生成建筑的形态。
尽管麦秸板作为结构未能实现，但它作为曾经的建筑节能、建筑低碳、
建筑更新的创作理念与形态，留给业界值得回味的印记。

Even though it has been seven years since Shanghai World Expo 2010, we designers never forget the explorations we have made for leaving China's unique footprints on the Expo history, especially the responsibilities that the engagement in the Vanke Pavilion project has awakened to us. The theme of the pavilion was global warming. Deeply recognized with the theme, we chose to base our architectural design on it. Especially given our conclusion of researches into the CO_2 emissions of buildings and building materials through their lifecycle and the fact that the pavilion would be a temporary venue, we creatively used straw board to build the facility and on such basis, worked out the appearance of the structure. Although the straw board no longer existed as a structure, it represented the energy-saving, low-carbon and renewed design ideas and did give the entire industry a thought-provoking impression.

多相工作室在2008年5月通过竞赛赢得万科馆的设计项目，之后经慎重选择，与北京金田建筑设计有限公司（后更名为“北京建院约翰马丁国际建筑设计有限公司”）、清华大学建筑设计研究院共同完成项目的建筑设计。之后历经两年，万科馆于2010年5月交付使用。我们在这里将讲述万科馆在建筑设计过程是如何生成的，对建筑节能的回应以及万科馆自身的建筑体验，我们并不把以上的各个方面看作是彼此割裂的，而是希望将其综合在一起得到解答。本文对于结构最初概念与最终实施方案的区别、机电设备设计、节能设计也进行了回顾。

和以往的经验不同的是项目主体设计团队第一次由三个设计团队联合组成，这无疑对于每一位设计人员来讲都是一次难忘的体验；同时万科的决策团队主要在深圳总部、项目的实施在上海，而主要的设计团队均在北京，对于项目的管理也是一种考验，作为上海世博会重要的企业馆之一，项目设计组还需要与多个展览设计团队协作来共同完成整个项目；项目的生态设计、照明设计、标识设计顾问的设计建议亦需在整个设计中得到落实和体现；第一次大规模使用的“麦秸秆板”也进行了连续的不间断的实验，同时有许多突破现行规范之处；在7年之后的今天，联合设计组认为，项目的完美实施依托于“合作的力量”。

一、万科馆的生成与实施

（一）建筑的生成

在万科馆设计竞赛阶段，万科馆的主题是“全球变暖”（万科馆最终确定的主题是“尊重的可能”，关注地球的生态与环境问题）。尽管当时展示内容还不明确，但基于对这一主题的认同，我们也将其作为了建筑设计的出发点。我们基于建筑及建材生命周期CO_2排放量的研究成果，并且考虑到万科馆是临时性场馆，在世博会闭幕后会被拆除（从建成到拆除仅有不到1年的时间），我们想尝试用秸秆板“盖房子”——将其作为建筑结构材料，然后以材料→结构→空间的推演为基础生成建筑的形态。万科馆在最终实施的时候结构方案已经改变，但建筑作为场所的独特要素，如空间、光线等依然存在，因此我们保留了最初的建筑形体与空间形态。

图10-1 方案鸟瞰图

秸秆是成熟农作物茎叶部分的总称。通常指小麦、水稻、玉米、油料、棉花、甘蔗等农作物在收获之后的剩余部分。农作物光合作用的产物有一半以上留存于秸秆当中。秸秆富含氮、磷、钾、钙、镁和有机质等，是一种具有多用途的可再生生物资源。秸秆也是一种粗饲料，其特点为粗纤维含量高（30%~40%），并含有木质素等。木质素虽然不能为猪、鸡所利用，但却能被反刍动物如牛、羊等所吸收。
目前，秸秆的用途主要有做饲料、还田、堆肥、发电、制建材等。
我国年产秸秆约 6 亿吨。

图 10-2 材料分析

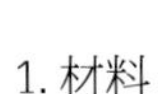

1. 材料

秸秆板是一种以质量上大约 95% 的秸秆和 5% 的异氰酸酯 (MDI) 胶，经热压成形制成的建材。秸秆通常指小麦、水稻、玉米、棉花、甘蔗等农作物在收获籽实后剩余的茎叶部分，农作物光合作用的产物有一半以上存在于秸秆中，由此制成的秸秆板就将农作物通过光合作用吸收的 CO_2 以有机碳的形式固化了下来。我们一开始关注秸秆板只是感觉它很亲切、对环境很“友好”，而对它更进一步的认识来自于委托清华大学建筑技术科学系提供的的一些研究成果和计算结果，这是用“生命周期评价”的方法对建筑及建材生命周期 CO_2 排放量做出的定量分析。

①根据《中国建筑环境影响的生命周期评价》对一栋典型板式多层住宅的研究结果，在建材生产（包括原材料生产）和建筑施工阶段消耗的总能耗中，建材生产的能耗占 92%，施工阶段的建材运输能耗占 5%，施工作业能耗占 3%。由此可见，选择什么材料来盖房子对于建筑生命周期的 CO_2 排放量至关重要。

②通过生命周期 CO_2 排放量分析，如果用秸秆板作为建材，1 千克秸秆板可固定 0.15 千克 CO_2。

③在秸秆板与各种常用建材的 1 千克材料生命周期 CO_2 排放量的比较中，秸秆板的优势非常明显。

2. 结构

如何用小块的片状的秸秆板组成包含着一定空间容积的结构？基于秸秆板具有一定的抗压强度、而抗拉强度与弹性模量很低的力学特性，在投标阶段，我们与结构工程师一起设计了一种结构形式——“加劲自平衡预压堆积体系”。这是一种类似于砌体结构的建造方式和结构体系：每层秸秆板之间靠销钉连接，上下层板错缝布置，并通过在高度方向上分段设置的水平钢板加劲层，与在竖向设置的预应力加劲绞线，将一块块的秸秆板连接为一个整体结构。对于这种结构形式，封闭的圆台是比较理想的稳定自平衡的形体，而多个不同圆台的组合也带来空间形态上的可能性，因此圆台就成为了构成空间的起点。

3. 空间

如何组织圆台？如果将正圆台（上小下大）与倒圆台（上大下小）作为两种基本的形体单元，它们的组合将获得三种基本空间形态：

①圆台内部空间—室内空间—展厅；

②倒圆台与倒圆台交接—空场空间—中庭；

③正圆台与倒圆台交接—夹缝空间—出入口；

4. 最终实施的结构与秸秆板的使用

由于万科方面觉得使用秸秆板叠涩这种结构方式在防火、造价方面存在风险，将结构改为了钢结构。时间的紧迫不允许我们在该项目里解决这些问题，我们感到非常遗憾。实施的万科馆是钢结构承重，外墙为多层复合构造，秸秆板不再承重，因此也就改为了披叠，作为建筑的外皮，可以起到防水的作用。

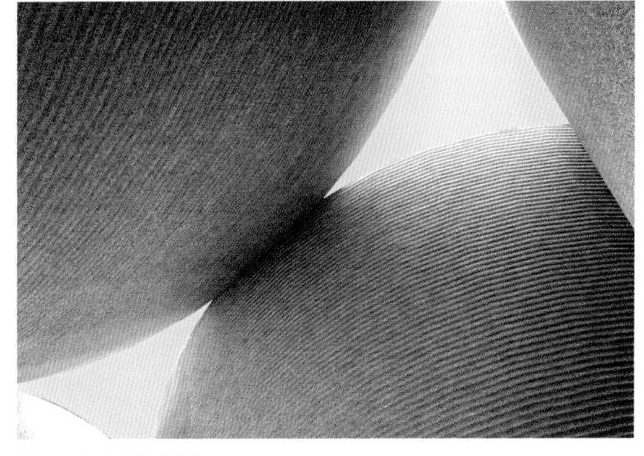
图 10-3 方案效果图

（二）空间

整个建筑的空间分为两种形态：圆台内部的空间与圆台外部的空间。圆台内部的空间是封闭的、静态的、单纯的，对应的功能是展厅、演播厅和后勤服务；而圆台外部的空间是开放的、流动的、复杂的，作为中庭功能主要是观众等候参观、问询和购买纪念品。

1. 对人的平衡感的挑战

倾斜的墙体不提供垂直参照，这对人的平衡感产生挑战，会使人上下左右地看这个建筑，以寻找并确认建筑的稳定。

2. 流动性

正、倒圆台的倾斜墙体彼此部分遮挡、部分展现，加之曲面的连续，使得中庭成为一个随着人的行动持续变化着的空间，让人产生探寻的渴望。

3. 开放性

世博会期间真正能进入万科馆内参观的人是少数的，因此我们希望形成一种可以和更多人分享万科馆的开放性的建筑。于是我们没有封闭中庭空间，并且用阻隔路径但不阻挡视线的水面围合建筑，而不是围栏、围墙，这样就可以使路上的行人不用进入万科馆就可以看到建筑内部，甚至看穿建筑。通过中庭顶部天窗和圆台之间缝隙的采光，“中庭平均天空光照度约为 500Lux”，使得中庭的景象可以从外部被看清。

图 10-4 方案效果图

（三）材料

1. 材料的时间性

新秸秆板的自然纹理和金黄色泽都会让人感受到生命的健康与丰盛，但如同任何生命会衰老死亡一样，秸秆板的色泽也会随着时间的推移而灰变。我们希望通过这种自然的褪变可以传达一个观念，即如果人们尊重自然的应有状态，就会减少与自然的无谓对抗。在设计过程中，一开始很多参与者都难以接受这种自然材料因日晒雨淋、时间的流逝而发生的褪色与灰变，他们认为必须去找到一种保护剂，让它在拆除时还能和新的一样，否则就要更换材料。不过他们最终都逐渐接受了这个大自然的决定。这件事挺有意思的，想想看，不会变旧是一件多么不自然的事情！但今天的人们却觉得这是再正常不过的。

2. 对触觉的激发

秸秆板的质感吸引了很多观众去触摸，因为这种不常见的材料超出了视觉固有的经验，他们需要触觉提供更多的信息，以确认这种材料。这种激发视觉之外的感官体验方面的设计（包括对人的平衡感的挑战）对我们来说非常重要，如果缺少就会将建筑变为单纯的视觉经验，那么体验建筑和看建筑照片之间的差别也就不大了。

（四）光

1. 自然光

7 个圆台及中庭顶部都留有 ETFE 充气膜采光天

图 10-5 总平面图

图 10-6 实景

图 10-7 实景

窗。自然采光不仅是为了节省能源，更重要的是，我们还想让建筑表现自然光之美：在晴天，阳光从顶部进入建筑，打在倾斜的墙面和地上，形成椭圆或曲线状的光斑，随着太阳的移动，光斑的位置、形状和亮度也在变化；在阴天或多云时，光线在建筑内部漫射；圆台的曲面也不会产生明确的明暗交界线……人们可以从中领略到各种表情的自然光带来的魅力。

2. 人工光

对于晚间的中庭，我们采用了"反射式照明"；而对于夜景照明，我们将设计的效果称之为"内透光"。

反射式照明是"从下部空间远距离向上投射并利用反射光为中庭提供功能照明"。我们在各圆台的门斗上部设计了藏灯的灯槽，利用倒圆台向地面倾斜的表面反射光线，实现了"见光不见灯"的照明效果。同时，反射色温为 3000K 的卤钨灯光线，可以很好地展现秸秆板的颜色和质感，营造了温暖的气氛。

从建筑外部透过圆台之间的缝隙，可以看到较外墙更为明亮的中庭，我们希望夜景照明可以加强这种明暗对比的效果，即"内透光"：建筑中庭的光线透过圆台之间的缝隙溢射出来。这样设计的原因在于：一是让光线好像从中庭溢出，从而使建筑内部空间获得一种富有吸引力的神秘感；二是照度最亮处在筒与筒之间缝隙，这样可以充分表达建筑由 7 个独立圆台

图10-8 东南侧实景

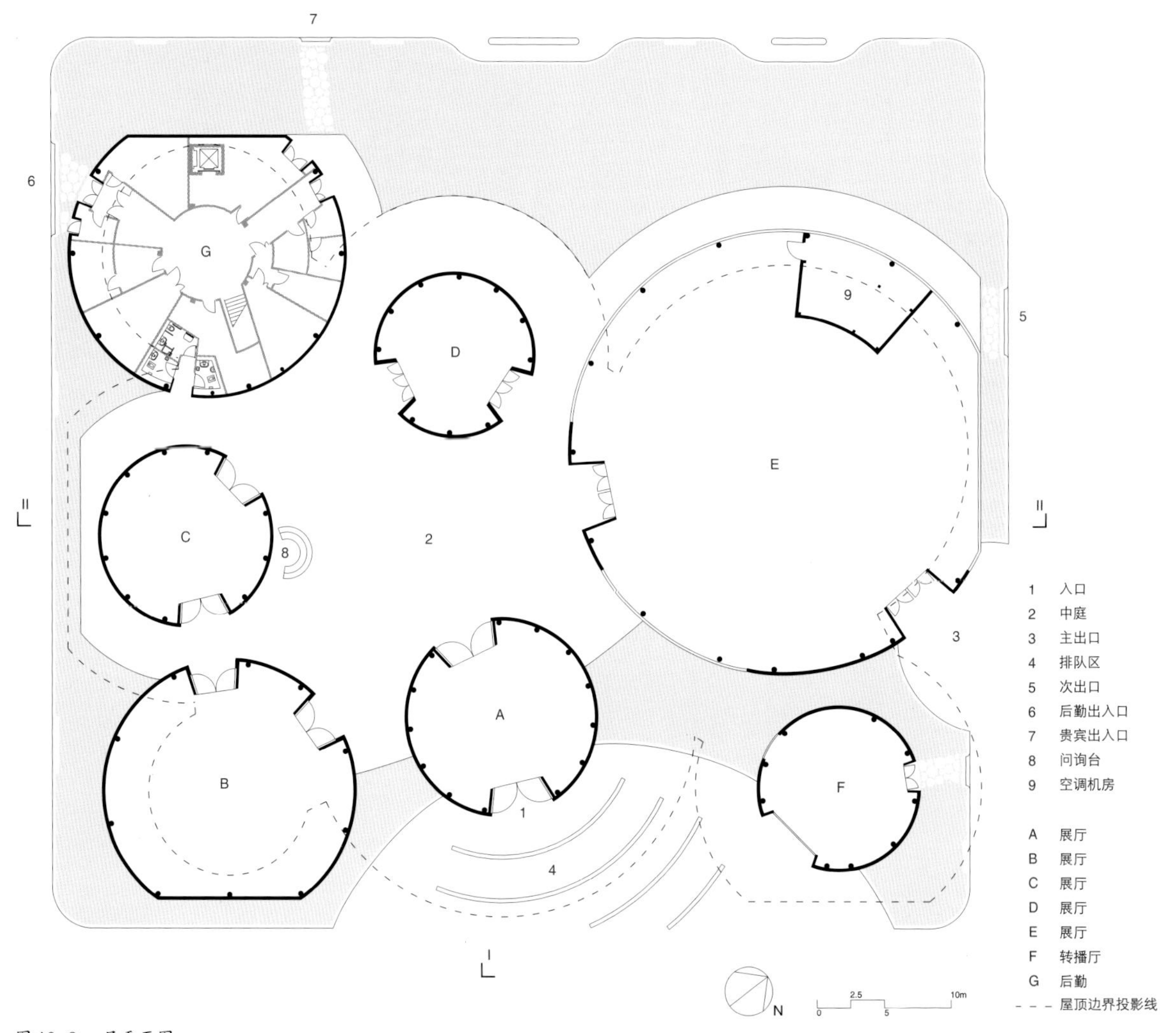

图 10-9 一层平面图

组成的形体关系；三是我们不选择将全部外立面照亮，这样可以使得用于外立面照明的能耗大为减少。最终夜景照明的实施，我们将绝大部分室外灯具的光都投向了筒与筒之间的缝隙，以加强中庭的光外溢的效果，事实上，这些光也增强了中庭的照度。个别几只灯具的光投向了圆台的切面，以显示切面的形状。

（五）水

围绕在建筑周边的水池提供建筑的倒影和必要的阻隔，同时，当室外空气温度超过 30 摄氏度和相对湿度低于 70% 的时候，水的蒸发可以降低进入室内空气的温度，这些水池的补水将使用从屋顶收集的雨水。

我们将花岗岩池边的设计符合座椅的高度和宽

度，使其成为对参观者友好的建筑边界。最终，大量的游人坐在万科馆低矮的池边歇脚、乘凉和戏水，在建筑形成的巨大落影下人群尤其密集，这与周边其他场馆采用围栏或建筑墙体来划定区域形成鲜明的对照。

（六）风

建筑中空调泛滥的结果，一方面建筑师已经越来越不会设计利用自然通风降温的建筑，另一方面使用者对自然通风降温越来越怀疑，风作为一个好的体验似乎在我们的建筑体验中逐渐消失了。我们想挑战一下这个让人"气闷"的现状：将占建筑总面积 1/6 的中庭设计为开敞的空间，完全不使用空调。中庭在 4 个方向均有开口，由于中庭周边的倒圆台形成的空隙都为上小下大，建筑外部的气流在经过建筑的时候，上部的气流被建筑的形体压向下部，从而使靠近地面的风速提高——风变大了，人会感到更凉爽。

中庭上部的 ETFE 膜气枕与女儿墙顶留有空隙，ETFE 膜在阳光照射下温度升高，可以加热中庭顶部空气，使顶部空气温度高于地面空气，实现热压通风。

虽然圆台内部由展陈设计负责，但我们还是预先在大多圆台的屋面安装了无动力自然通风器（涡轮通风器），希望这些通风器在适宜的时间（温和季节或早晚时段）靠自然风无动力运行，利用自然风力抽出室内空气，而在使用空调时通过电动风阀封闭。各圆筒中央均设有电动高侧窗，可在适宜的时间开启，利用屋顶与地面的温度差实现自然通风。

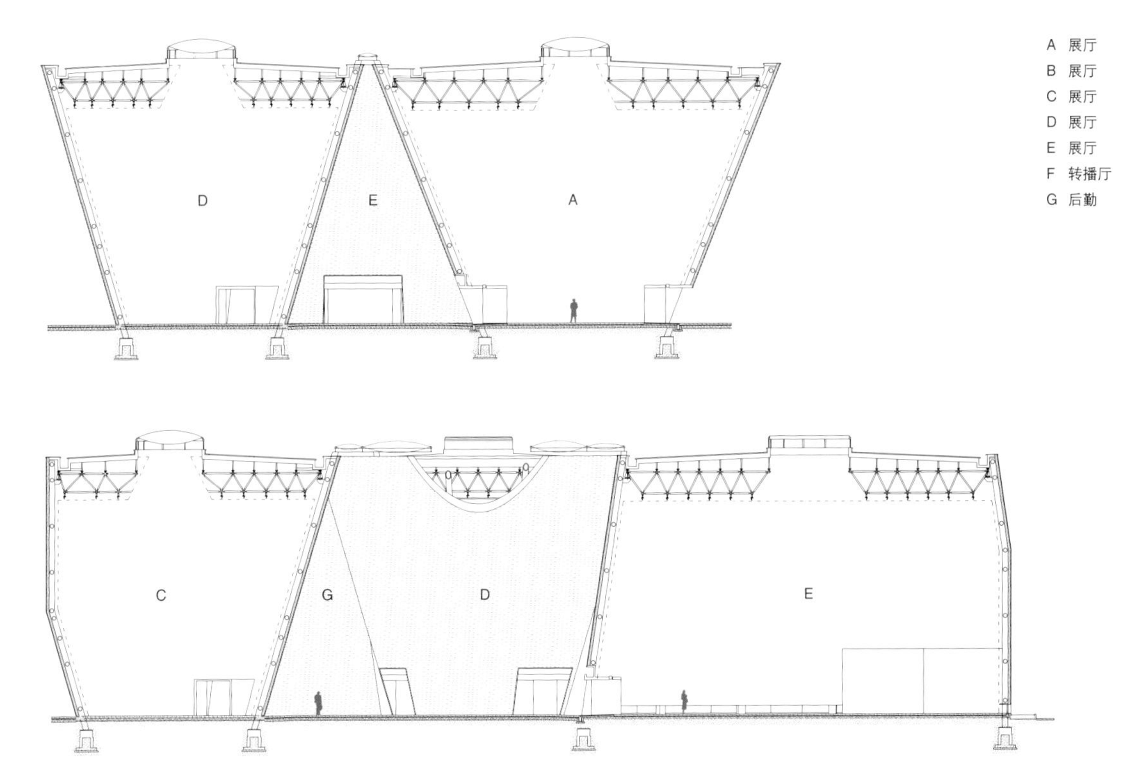

图 10-10 剖面图

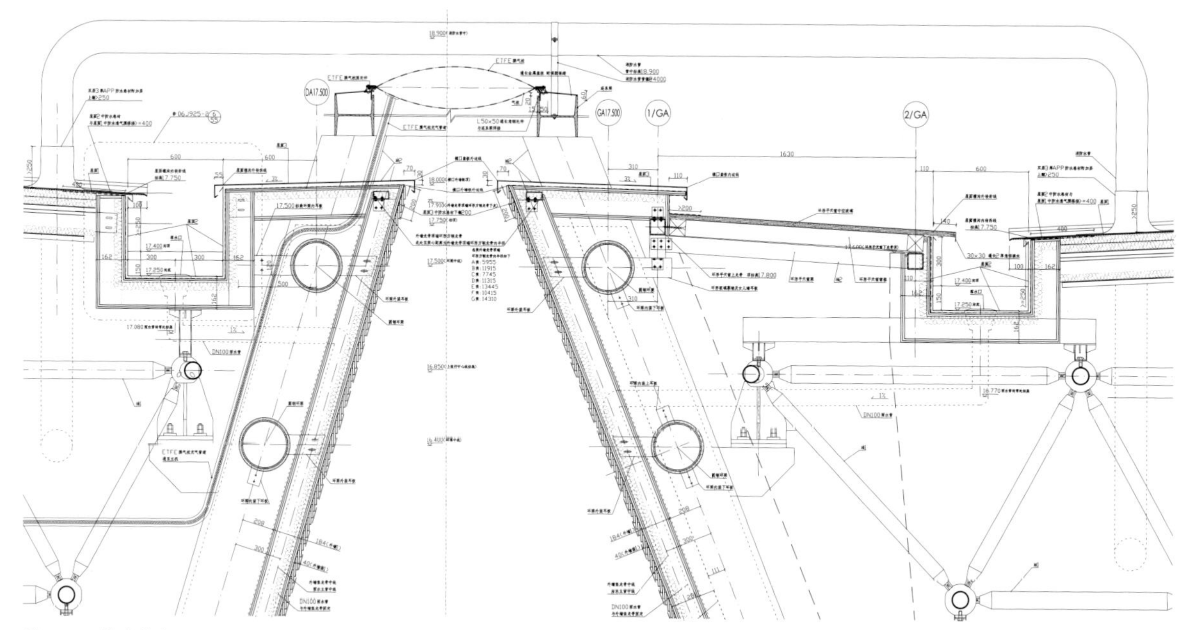

图 10-11 构造节点

二、万科馆的结构

在实施方案中，综合各方的考虑，主体结构采用了钢结构，秸秆板作为维护材料，仅在入口接待台局部按照投标方案构想进行了建造（已经显示了其独特和有趣的营造能力，我们也会在今后的工程中继续完善推进这种新体系的实践）万科馆主体结构外墙为秸秆板披叠构造，秸秆板用 2 颗沉头自攻螺丝固定，上下层秸秆板错缝。在窗户及进出风口处，采用秸秆板百叶；秸杆板百叶的内侧设置铝合金的弧形龙骨及托片，秸秆板与托片连接。部分筒的外墙体采用了上下层板之间增加垫块叠压的方式。为了减轻自重，屋顶则依旧保留了投标阶段的 ETFE 充气膜结构，并利用筒体间的连系梁形成架空拔风构造。

（一）主体结构方案

结构主体包括 7 个大小不等的圆锥筒钢结构，最大直径约 32 米。

圆锥筒外壁主体结构采用钢管斜柱与弧形钢管梁刚性相贯连接体系，圆锥筒内部及楼面结构采用钢管柱与 H 形钢梁组成框架体系。

部分环梁根据建筑开门需要断开，并在局部进行了抽柱。设计中利用标高 5.4 米环梁与 3.6 米环梁分别作为上下弦设置了转换桁架。屋面采用四角锥网架与铺设金属屋面板的形式。网架支座底面标高 16.4 米，上弦多点支承。屋顶小立柱找坡。中庭采光顶采用 ETFE 充气结构。筒体间通过联系梁（标高 18.3 米）在屋面连接成整体。在有联系梁处网架上弦与顶环梁间形成顶桁架，用来加强筒体之间的连接。

（二）节点构造

本工程采用了圆管梁柱体系，由于梁为弧形，在柱间设置支撑会影响墙面做法。梁柱节点需要做成刚接。设计中采用了圆管相贯并增加竖向插板和水平环

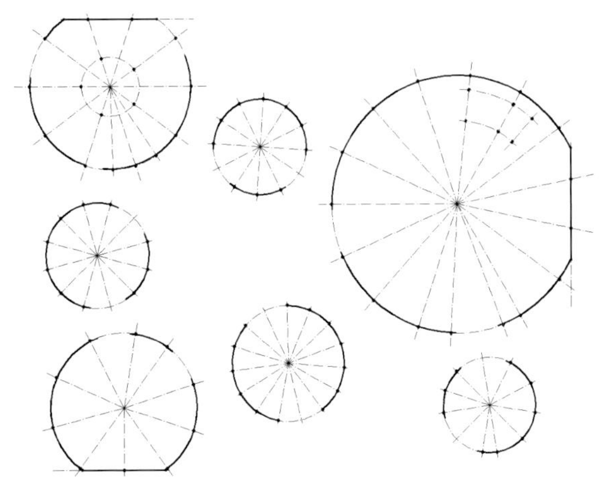
图 10-12 地面附近环梁平面

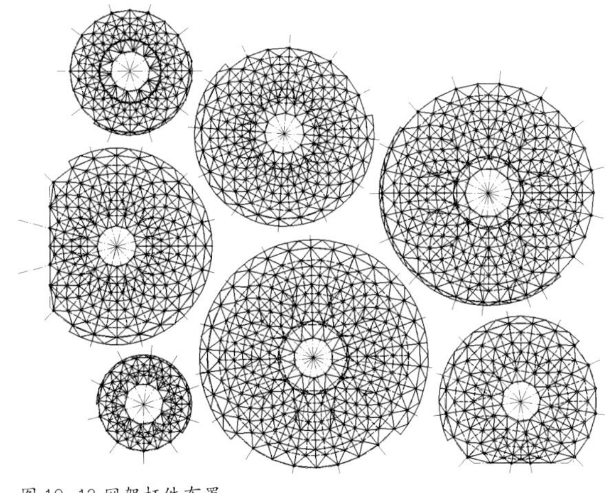
图 10-13 网架杆件布置

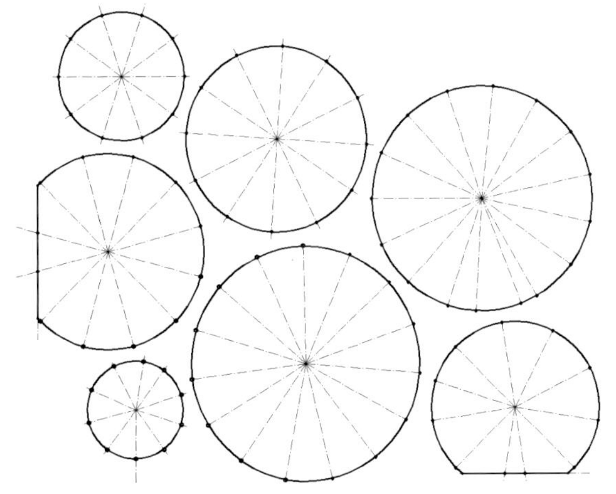
图 10-14 屋顶环梁（标高 17.5 米）平面

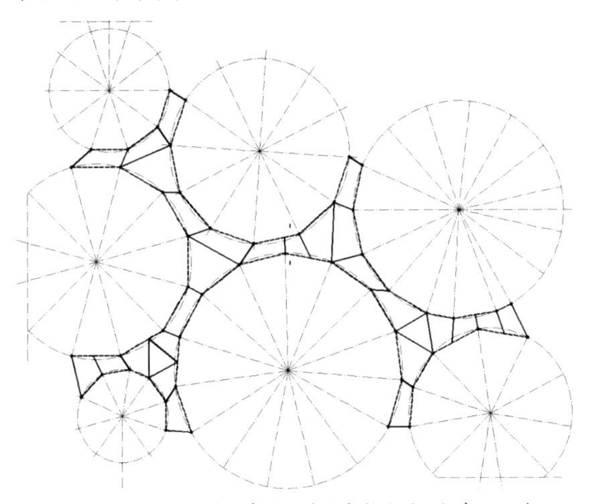
图 10-15 筒体间通过联系梁在屋面连接成整体图，标高 18.3 米

板的构造，满足节点处的抗弯需要。

虽然实际工程中没有采用秸秆作为结构材料的方案，但作为结构工程师，积极投入应对全球变暖的低碳设计的努力仍是值得尝试的。同时，新材料的使用也自然会激发出完全不同于传统的结构形式，期待由此产生的越来越多的全新设计能不断推进我国工程建设的持续发展。

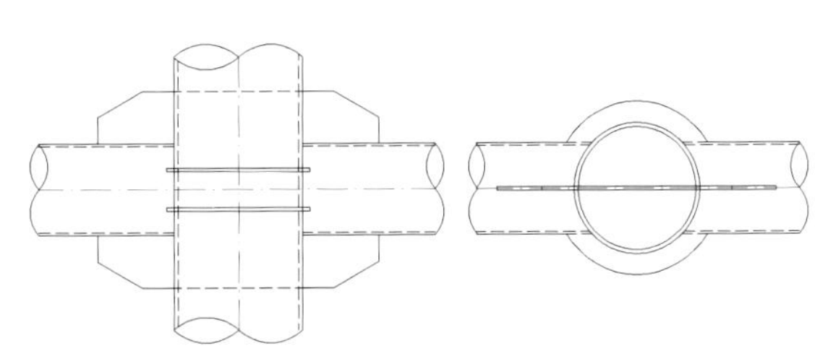
图 10-16 结构节点

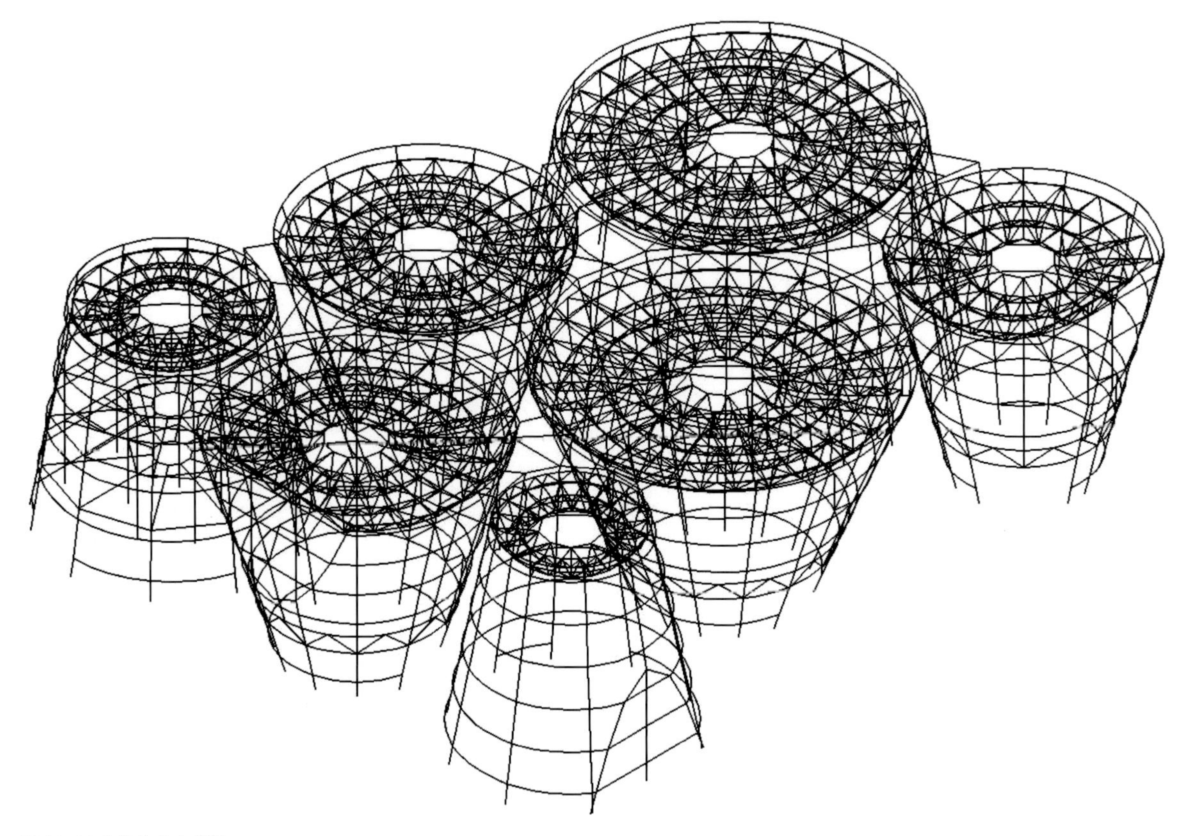

图 10-17 结构体系分析图

三、万科馆的机电设备

上海世博会万科馆建筑面积 5000 平方米，建筑高度 19.5 米，为世博会期间临时展馆，使用年限 1 年。

（一）空调设计

空调冷源由街坊集中冷冻机房提供 6.5/13.0 摄氏度冷冻水，冷冻水系统工作压力 0.6 兆帕，可资用压头 0.18 兆帕，冷量 2000 千瓦，接入管管径 DN250。空调水系统采用两管制，用于夏季供冷。根据建筑使用功能及空间结构形式，分别采用风机盘管、风机盘管加新风及一次回风全空气空调系统：等候厅、前展厅、后展厅及 VIP 贵宾接待等区域利用风机盘管，采用分层空调系统形式夏季供冷，即空调冷风只送至人员活动区域，各厅口处设置风幕；过渡季节利用门洞、屋顶可开启天窗及桶状结构的吹拔作用自然通风；观影主展区设置一次回风全空气空调系统，采用吊顶下旋流风口送风，观众及表演区回风方式。

（二）给排水设计

展馆内生活用水由室外市政给水管网直接供给，供水压力 0.2 兆帕，供水能力 183 立方米 / 日，接入管管径 DN100；室外水景用水由给水管网补给，补水处设置倒流防止器；室内生活污、废水分流，排至室外污水窨井,经区域化粪池处理后,排入市政污水管网;展馆屋面雨水经内排水管道收集后排至室外;绿地雨水回渗土壤;道路采用透水率高的铺装材料;径表水及水景溢流水排至室外市政雨水管网。室外排队等候区域设置水喷雾系统,为参观顾客降温。

工程内设置有室外消火栓、室内设消火栓给水系统及自动喷洒系统，其中：室外消火栓用水量 30 升 / 秒；室内消火栓用水量 20 升 / 秒；自动喷洒灭火系统按照中危险 I 级设计，用水量 20.8 升 / 秒；大空间洒水灭火装置用水量 47 升 / 秒。室内外消防用水由世博会提供，消防水系统分两路供给，接入管管径 DN250；供水压力 0.7 兆帕。室内外消防系统均为环状管网，并设置水泵接合器。室内净空高度≥ 12 米中庭等设置大空间洒水（水炮）灭火装置。无自然排烟条件的内走道及高度大于 12 米的中庭等设消防机械排烟系统。灭火器按照严重危险级 A 类火灾设计，配置基准 50 平方米 /A。

图 10-18 水体实景

（三）电气设计

本工程设有供配电系统、照明系统、动力系统、安防系统、火灾自动报警及消防联动系统、综合布线系统（因无具体工艺资料，本设计仅预留条件）。其中展示工艺本设计仅在设计中预留电源至工艺配电盘，配电盘后接线由工艺施工方自行解决。主展厅、分展厅、后展厅展示用电本设计仅预留电源条件其余由电通公司深化完善设计。

1. 强电设计

企业馆因为本身负荷较大由市政电力外线提供十路 380/220 伏电缆（因本工程为低压用电户，所有低压外线由园区箱式变电站引来，其中本馆内的一、二类负荷应由不同高压回路的箱式变电站引来，根据负荷计算容量按 1800 千伏安考虑。

低压系统（380/220 伏）考虑低压电缆截面不宜过大采用十路低压电缆分别引入，其中分为：主展厅演出工艺 6 路；其他展厅演出工艺 2 路；馆内其他电力负荷 2 路。其中应急电源应双电源、双回路一用一备在负荷侧实行切换，当任何一侧非事故失电时，低压双电源转换开关（ATSE）采用自投自复方式（应设机电闭锁装置，防止误操作，确保两路电源不并列运行）。本工程根据世博会要求按二类防雷设置防雷系统。在企业馆屋顶设置预作用式避雷针与建筑物钢结构连接后引至室外综合接地体。

2. 弱电系统设计

工程设计了综合布线系统、有线电视系统、建筑设备监控管理系统（实现对风机盘管、室内机的统一管理）、安全防范系统（闭路监控、出入口管理、入侵报警、电子巡更、安检）、信息发布系统（仅预留条件）、消防及广播系统等。

弱电机房应设置于建筑物首层。本机房包括网络、市话、移动、网通的主干进线设备及 UPS 电池组。消防控制中心兼作安防中心、电视机房、信息发布机房、建筑设备监控管理、广播系统机房。消防中心应有直接对室外开放的门,同时尽量远离设备专业用房。

图 10-19 圆台缝隙实景

机房内除设置各类管理主机及相关设备外，还应设置UPS 电池组。消防中心设置在首层，且室内环境应满足设备运行条件。

综合布线系统由六个子系统组成：工作区子系统、水平区子系统、管理子系统、主干线子系统、设备间子系统、建筑群子系统。

本工程安全防范系统包括如下子系统：①闭路监控系统；②出入口管理系统；③入侵报警系统；④电子巡更系统；⑤安检系统。

本建筑物在建筑物首层设有消防控制中心，内部除卫生间以外的所有部位均为报警监控范围，整个建筑按二级防火等级的要求进行消防设计。系统形式为控制中心报警系统，火灾报警回路采用二总线方式，高大空间采用红外对射式火灾探测器。

消防广播与背景音乐公用一套扬声器，平时作为背景音乐使用，消防时由消防主机强切作为消防紧急广播。扬声器在场馆室外、场馆内公共区域、走廊内的设置。系统的音源、功放、主机等控制设备设置在消防控制中心。

在大堂、办公区、展区、走道等公共区域设置背景音乐及公共广播扬声器用以播送背景音乐，每个区域设置独立的音量控制器，当遇到紧急事故广播时，不管音量控制器的开关是否关闭，均能自动切换到紧急事故广播上，以最大音量播送紧急事故广播。

（四）绿色建筑设计

在万科馆的设计、实施过程中，践行绿色、环保、节能理念，外檐维护结构保温按照节能75% 标准设计；风机盘管、空调器根据计算冷负荷合理选用，保障空调设备在高效区间运行；高大空间采用分层供冷空调形式，空调冷风只送至人员活动区域；分区域设置自动温控系统；各展厅入口设计自动风幕气帘，有效降低室内冷量损失；利用门洞、屋顶可开启天窗及桶状结构的吹拔作用自然通风……在世博技术委员会及万科组织的专家委员会审核评审中获得“机电设备专业未见可优化处”的评价。

注释

[1] 部分内容参考百度百科（http://baike.baidu.com/），“秸秆”词条，部分数据来源：江苏大盛板业有限公司

[2] 生命周期评价：Life Cycle Assessment（LCA），是目前国际上公认的产品系统环境负荷的量化评价方法。联合国环境规划署（UNEP）对于LCA的定义为：“LCA是评价一个产品系统生命周期整个阶段——从原材料的提取和加工，到产品生产、包装、市场营销、使用、再使用和产品维护，直至再循环和最终废物处置——的环境影响的工具。”生命周期是一个从摇篮到坟墓的过程。

[3] CO_2排放计算说明：秸秆板重量的5%为MDI胶，95%为秸秆，采用热压工艺，成品密度650kg/m^3，生产1m^{3}秸秆板耗电350kWh；秸秆含碳率40%，江苏小麦单产6041kg/hm^2，小麦经济系数0.4，农田机械管理耗油78kg/hm^2，施肥362kg/hm^2，小麦秸秆还田率45.5%；原材料生产阶段各种运输过程的CO_2排放已计入原材料生产阶段的总排放中。部分计算基于清华大学建筑技术科学系顾道金博士的研究成果及其开发的配套分析软件（Building Environmental Load Evaluation System, BELES），除秸秆板材外，其余建材不包括施工阶段的CO_2排放。

项目信息及版权所有

世博会万科企业馆
业主：万科企业股份有限公司
建设地点：上海世博园E片区（浦西，半淞园路/西藏南路）
建筑设计：多相工作室+北京建院约翰马丁国际建筑设计有限公司+清华大学建筑设计研究院有限公司
建筑及景观方案设计：多相工作室
建筑及景观建筑专业施工图设计：多相工作室、北京建院约翰马丁国际建筑设计有限公司
建筑及景观设备、电气专业施工图设计：北京建院约翰马丁国际建筑设计有限公司
建筑结构专业方案及施工图设计：清华大学建筑设计研究院
建筑主体设计团队：1.多相工作室 2.朱颖 3.常强 4.谷凯 5.熊进华 6.姜建中 7.薛磊
建筑生态技术咨询：1.林波荣 2.谷立静 3.周潇儒 4.彭渤（清华大学建筑学院建筑技术科学系）
建筑照明设计咨询：张昕（清华大学建筑学院建筑技术科学系）
指示系统设计：广煜（吐毛球平面设计）
展览设计及展厅室内(A~F筒)设计：北京盛阳世纪文化传播有限公司、北京华毅司马展览工程有限公司、北京龙安华诚建筑设计有限公司
基地面积：5000平方米 总建筑面积：3009平方米
项目状态:已建成
设计时间：2009年 建成时间:2010年

撰文：万科馆的建筑生成与实施/多相工作室+清华建筑设计研究院+建院马丁万科馆联合设计组
（一）万科馆的建筑生成与实施/多相工作室
多相工作室（duoxiang studio）于2006年10月在北京成立，合伙人为陈龙、胡宪、贾莲娜、陆翔（人名按姓氏拼音首字母排序）。
（二）万科馆的结构/清华大学建筑设计研究院有限公司运算化设计国际研究中心主任 常强
（三）统稿与万科馆的机电设备/北京建院约翰马丁国际建筑设计有限公司 朱颖 谷凯 姜建中

附录一　书法：马到成功

Appendix I　Calligraphy: Madao-Chenggong

著名建筑师、艺术家朱小地先生为建院马丁公司改组七周年题写“马到成功”
2014 年

馬到成功
甲午年冬 小地

附录二　书法：马顶（丁）羊肥

Appendix II　Calligraphy: Mading-Yangfei

全国工程勘察设计大师何玉如先生录全国工程勘察设计大师刘力先生为建院马丁公司羊年新春贺词“马顶（丁）羊肥”
2015 年

頂
肥
录劉力大師為建院馬丁公司羊年新春賀詞玉如書

附录三 书法：吉安文化艺术中心赋

Appendix III Calligraphy: The Fu of Ji'an Cultural and Art Center

全国工程勘察设计大师何玉如先生摘录著名作家
赖卫东先生创作《吉安文化艺术中心赋》
2015 年

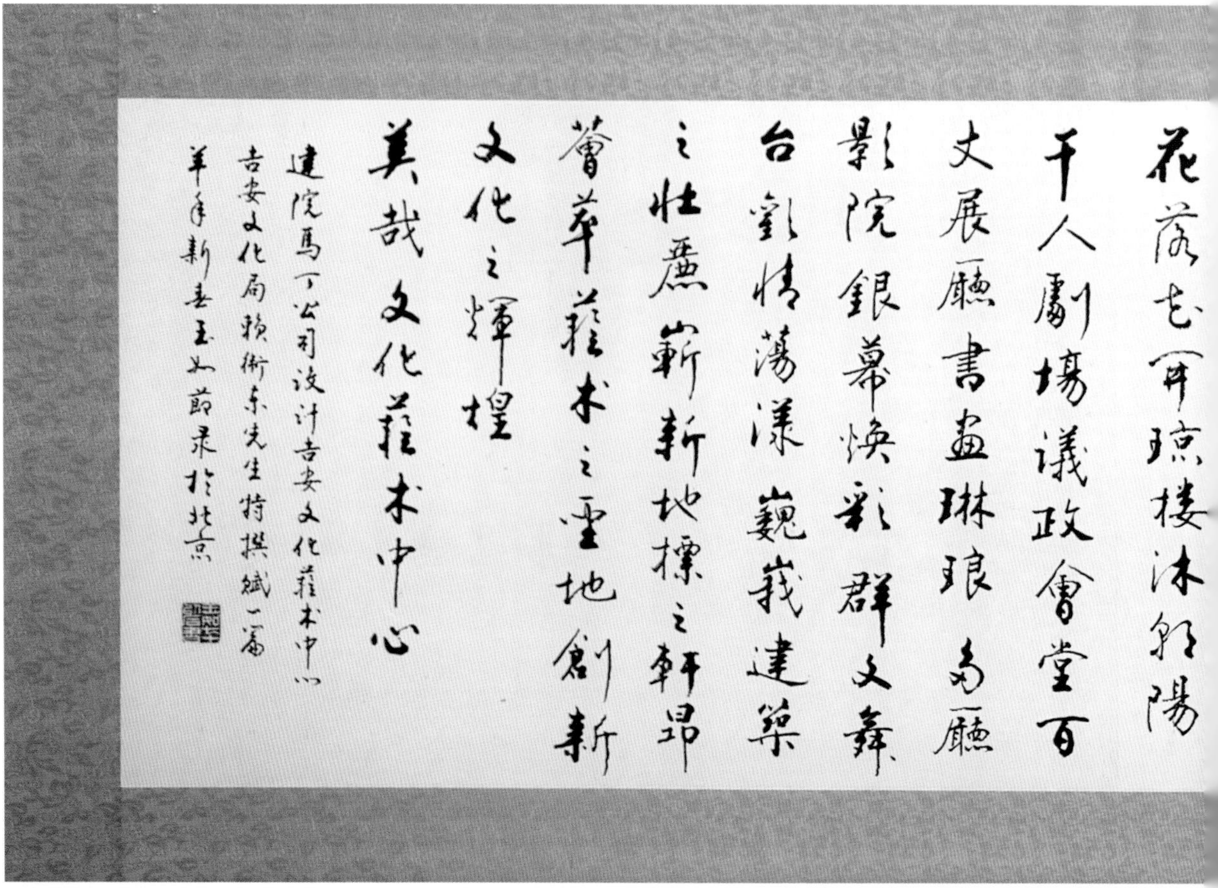

吉安文化艺术中心赋

作者：赖卫东

文化吉安，庚寅添彩。城南新建文化艺术中心，驻足凝眸，遐思畅想。喷泉舞，卿云升，霓虹闪，幻影动。疑是天宫落凡尘，如闻天籁动琴心。

吾乡庐陵，风流倜傥。五千年文化积淀，博大精深；三十载改革开放，星高云灿。古今灵秀之地，聚紫气而奋发；文章节义之邦，著新篇而图强。经济勃兴，社会昌明，江山多娇，春潮浩荡。君不见：园区龙腾虎跃，高潮叠浪千舟发；城区云蒸霞蔚，锦绣辉煌万众裁。百鸟在绿荫中翻飞，文明在繁荣中升华。广场腾舞，社区欢歌，城乡如画，百姓安康。

发展呼唤文化引航，时代期盼艺术繁昌。构和谐，顺民心，重文化，筑殿堂。云卷云舒，宏图变美景；花落花开，琼楼沐朝阳。千人剧场，议政会堂；百丈展厅，书画琳琅。多厅影院，银幕焕彩；群文舞台，欢情荡漾。巍峨建筑之壮丽，崭新地标之轩昂，荟萃艺术之圣地，创新文化之辉煌。

美哉！文化艺术中心！以小见大，以近思远。文化兴则民族旺，大众乐则天下康。愿文化之树常青，艺术之花常开。无限前途，无限作为，无限美景，无限遐想！

壮哉！吉安！风华正茂，磅礴激昂！

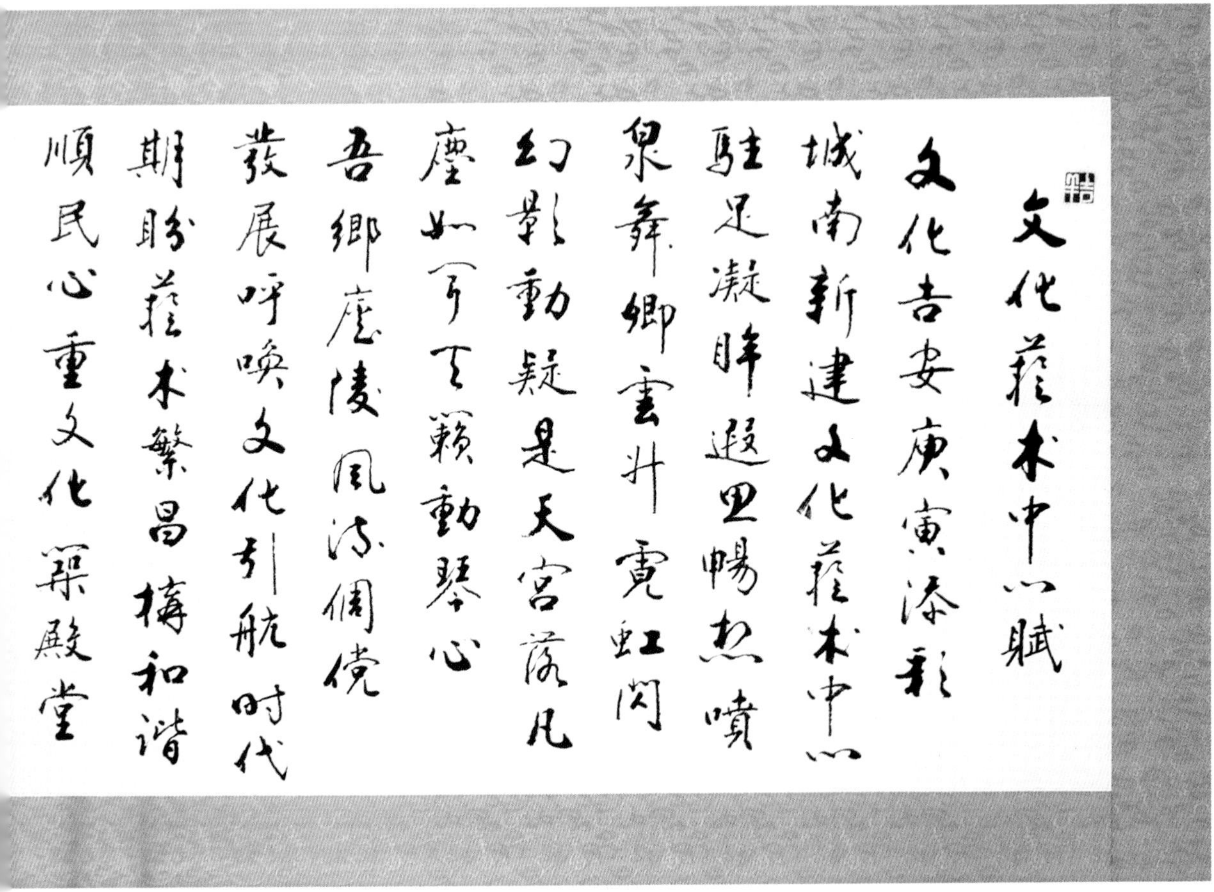

附录四
获奖作品一览

Appendix IV
Honored Works

序号	项目名称	获奖等级	颁发部门	年份	主要获奖人
1	江西吉安文化艺术中心	首都第十七届城市规划及建筑设计方案汇报展优秀方案奖	北京市规划委员会	2010	朱颖、邹雪红、王鹏、朱琳、沈桢、韩涛、葛亚萍、苗启松、徐斌、张胜、周宏宇、谷凯、熊进华、姜建中、薛磊
		“江西省十佳建筑”称号	江西省政府	2012	
		2014年度BIAD优秀工程设计奖公共建筑类一等奖	北京市建筑设计研究院有限公司	2014	
		北京市优秀建筑工程设计项目二等奖	北京市勘察设计协会	2015	
		全国优秀工程勘察设计行业奖二等奖	中国勘察设计协会	2015	
2	西安园博会创意自然馆及园区大门	首都第十七届城市规划及建筑设计方案汇报展优秀方案奖	北京市规划委员会	2010	项目设计组
3	晋国博物馆	首都第十七届城市规划及建筑设计方案汇报展优秀方案奖	北京市规划委员会	2010	项目设计组
4	河南省委党校规划	首都第十七届城市规划及建筑设计方案汇报展优秀方案奖	北京市规划委员会	2010	项目设计组
5	回龙观文化居住区F05区工程北区设计	北京市2011年保障房汇报展保障房十佳设计	北京市规划委员会	2011	项目设计组

续表

序号	项目名称	获奖等级	颁发部门	年份	主要获奖人
6	华北油田任丘石油工业遗址公园	首都第十八届城市规划及建筑设计方案汇报展优秀方案奖	北京市规划委员会	2011	项目设计组
7	昌平巩华城北区回迁住宅B区	首都第十八届城市规划及建筑设计方案汇报展优秀方案奖	北京市规划委员会	2011	项目设计组
8	昌平区回龙观A08地块配套商业用房项目	2012年度“BIAD设计”杯优秀工程设计（公共建筑）二等奖	北京市建筑设计研究院有限公司	2012	朱颖、沈桢、王鹏、邹雪红、曾劲、朱琳、田玉香、董小海、王琼、熊进华、姜建中、彭晓佳、薛磊、许阳、赵静远
		北京市建筑工程优秀设计奖	北京市勘察设计协会	2015	
9	某科研中心	2014年度BIAD优秀工程设计奖公共建筑类二等奖	北京市建筑设计研究院有限公司	2014	朱颖、邹雪红、王鹏、朱琳、葛亚萍、田玉香、周宏宇、许阳、熊进华、赵伟、张金玉、姜雅卉、姜建中、薛磊
		北京市建筑工程优秀设计奖	北京市勘察设计协会	2015	
10	顺义区李桥镇商业金融办公区（翼之城）	2014年度BIAD优秀工程设计奖公共建筑类三等奖	北京市建筑设计研究院有限公司	2014	朱颖、张彤梅、杨林、胡益莎、田玉香、王越、王琼、闫志雄、姜雅卉、彭晓佳、姜建中、郭王伟
		北京市建筑工程优秀设计奖	北京市勘察设计协会	2015	
11	潍坊滨海经济开发区城市展览馆	第十一届中国钢结构金奖		2014	项目设计组

续表

序号	项目名称	获奖等级	颁发部门	年份	主要获奖人
12	平谷区第一职业学校综合实验楼	北京市优秀建筑工程设计项目——抗震防灾专项奖	北京市勘察设计协会	2015	苗启松、陈晗、马培培、阁东东、李文峰、卢清刚、范波、张胜
13	通用电气医疗中国研发试产运营科技园（GE）	2017年度BIAD优秀工程设计奖公共建筑类一等奖	北京市建筑设计研究院有限公司	2017	邹雪红、朱颖、葛亚萍、鲁晟、朱琳、周彰青、张涛、田玉香、张胜、章伟、常青、赵伟、彭晓佳、李轩、刘昕、杨一萍
		北京市优秀建筑工程设计项目二等奖	北京市勘察设计协会	2017	
		全国优秀工程勘察设计行业奖一等奖	中国勘察设计协会	2017	
14	又见五台山剧院	2015年度BIAD优秀工程设计奖公共建筑类一等奖	北京市建筑设计研究院有限公司	2015	朱小地、高博、朱颖、罗文、田立宗、孔繁锦、贾琦、韩涛、田玉香、赵伟、赵阳、王越、张胜、章伟、姜雅卉
		2016中国建筑学会建筑创作奖金奖（公共建筑类）	中国建筑学会建筑师分会	2016	
		亚洲建筑师协会公共建筑设计提名奖	亚洲建筑师协会	2016	
		北京市优秀建筑工程设计项目一等奖	北京市勘察设计协会	2017	
		全国优秀工程勘察设计行业奖一等奖	中国勘察设计协会	2017	
		中国建筑学会建筑创作奖	中国建筑学会	2017	

附录五
主要作品名录

Appendix V
The Main Works Directory

内蒙翔宇盛乐新城二期别墅
建筑面积：33757 平方米
项目位置：内蒙古自治区呼和浩特市
设计年份：2008 年

上海世博会沙特馆设计竞赛
建筑面积：5300 平方米
项目位置：上海市
设计年份：2008 年

上海世博会铁路馆设计竞赛
建筑面积：5600 平方米
项目位置：上海市
设计年份：2008 年

上海世博会匈牙利馆
建筑面积：5100 平方米
项目位置：上海市
设计年份：2008 年
合作机构：TAMAS LEVAI

北京工具厂经济适用住房项目
建筑面积：230000 平方米
项目位置：北京市通州区
设计年份：2008 年

北京昌平区老城核心区城市设计
规划面积：200 公顷
建筑面积：1800000 平方米
项目位置：北京市昌平区
设计年份：2009 年
合作机构：MAKE（美国）

大连港 15 库创意中心
建筑面积：27000 平方米
项目位置：辽宁省大连市
设计年份：2008 年

吉安文化艺术中心
建筑面积：22500 平方米
项目位置：江西省吉安市
设计年份：2009 年
合作机构：北京市建筑设计研究院有限公司
复杂结构研究院

北京轨道交通昌平线一期、二期 TOD 研究
规划面积：90 公顷
项目位置：北京市昌平区
设计年份：2009 年

上海世博会万科馆
建筑面积：4400 平方米
项目位置：上海市
设计年份：2008 年
合作机构：多相工作室、清华大学建筑设计院

河南省委党校规划设计竞赛
建筑面积：297740 平方米
项目位置：河南省郑州市
设计年份：2009 年

北京市顺义区李桥镇商业金融（南侧）项目
建筑面积：152860 平方米
项目位置：北京市顺义区
设计年份：2010 年
合作机构：UA 国际

昌平区回龙观 A08 地块配套商业用房项目
建筑面积：27023 平方米
项目位置：北京市昌平区
设计年份：2010 年

金隅万科城幼儿园
建筑面积：3627 平方米
项目位置：北京市昌平区
设计年份：2010 年

北京市棉麻公司综合业务楼
建筑面积：16395 平方米
项目位置：北京市丰台区
设计年份：2011 年

北京国际旅行卫生保健中心综合楼
建筑面积：15115 平方米
项目位置：北京市东城区
设计年份：2010 年

北京金隅万科城项目 A2~A5\A7 地块
建筑面积：84136 平方米
项目位置：北京市昌平区
设计年份：2010 年

2011 西安世界园艺博览会标志性建筑设计
基底面积：25000 平方米
项目位置：陕西省西安市
设计年份：2011 年
合作机构：PLASMA STUDIO（UK）

昌平区人口和计划生育综合服务中心
建筑面积：7259 平方米
项目位置：北京市昌平区
设计年份：2010 年

某科研中心
建筑面积：5730 平方米
设计年份：2011 年

首邑溪谷
建筑面积：250000 平方米
项目位置：北京市大兴区
设计年份：2011 年
合作机构：北京市建筑设计研究院有限公司
4A1 工作室

首创伊林郡
建筑面积：100000 平方米
项目位置：北京市房山区
设计年份：2011 年
合作机构：北京市建筑设计研究院有限公司
4A1 工作室

青岛体育中心体育场改造
建筑面积：84193 平方米
项目位置：山东省青岛市
设计年份：2012 年
合作机构：北京市建筑设计研究院有限公司
第二设计院

郑州百荣 CSD 商贸区总体规划
建筑面积：4000000 平方米
项目位置：河南省郑州市
设计年份：2012 年
合作机构：河南省建筑设计研究院

北京巩华城农民回迁保障性住房
建筑面积：699830 平方米
项目位置：北京市昌平区
设计年份：2011 年

布拉柴维尔体育中心设计竞赛
建筑面积：333939 平方米
项目位置：刚果共和国布拉柴维尔
设计年份：2012 年

柳州银泰新银都中心
建筑面积：91099 平方米
项目位置：广西省柳州市
设计年份：2012 年
合作机构：和桥设计

北京开关厂定向安置房项目
建筑面积：201100 平方米
项目位置：北京市丰台区
设计年份：2011 年

潍坊滨海城市展览馆
建筑面积：20000 平方米
项目位置：山东省潍坊市
设计年份：2012 年
合作机构：朱培建筑设计事务所

霍州煤电集团安全技术教育培训中心
建筑面积：141205 平方米
项目位置：山西省临汾市
设计年份：2012 年

通用电气医疗中国研发试产运营科技园
建筑面积：74200 平方米
项目位置：北京亦庄经济技术开发区
设计年份：2013 年

郑州百荣希尔顿逸林酒店及甲级写字楼
建筑面积：155536 平方米
项目位置：河南省郑州市
设计年份：2013 年
合作机构：河南省建筑设计研究院

河北冀州铜锣湾广场
建筑面积：255400 平方米
项目位置：河北省冀州市
设计年份：2013 年

又见五台山剧院
建筑面积：13949 平方米
项目位置：山西省忻州市
设计年份：2013 年
合作机构：北京市建筑设计研究院有限公司艺术中心、第七设计院等

郑州百荣 CSD 商贸城一期
建筑面积：510000 平方米
项目位置：河南省郑州市
设计年份：2013 年
合作机构：河南省建筑设计研究院

大厂华夏幸福学校
建筑面积：53670 平方米
项目位置：河北省廊坊市
设计年份：2013 年

民生艺术中心（原世博会法国馆改造）
项目建筑面积：9060 平方米
项目位置：上海市
设计年份：2013 年
合作机构：朱培建筑设计事务所

郑州百荣 CSD 商贸城二期
建筑面积：1320000 平方米
项目位置：河南省郑州市
设计年份：2013 年
合作机构：河南省建筑设计研究院

首开国风美唐
建筑面积：551562 平方米
项目位置：北京市昌平区
设计年份：2013 年
合作机构：VP（法国）

沙河镇七里渠南、北村定向安置房工程
建筑面积：312643 平方米
项目位置：北京市昌平区
设计年份：2013 年

山西运城河东文化产业园概念性城市设计
规划面积：368 公顷
项目位置：山西省运城市
设计年份：2013 年

定州体育中心一期工程设计
建筑面积：61000 平方米
项目位置：河北省定州市
设计年份：2014 年

北京昌平区东扩新区控制性详细规划
规划面积：1140 公顷
项目位置：北京市昌平区
设计年份：2013 年

石家庄鹿泉文化新城城市设计
规划面积：900 公顷
项目位置：河北省石家庄市
设计年份：2013 年

北京民生美术馆
建筑面积：32900 平方米
项目位置：北京市朝阳区
设计年份：2014 年
合作机构：朱培建筑设计事务所

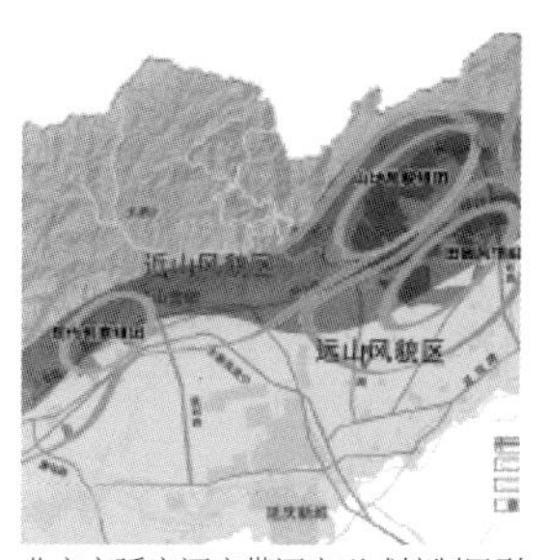

北京市延庆酒庄带酒庄形式控制导引
规划面积：10000 公顷
项目位置：北京市延庆区
设计年份：2013 年

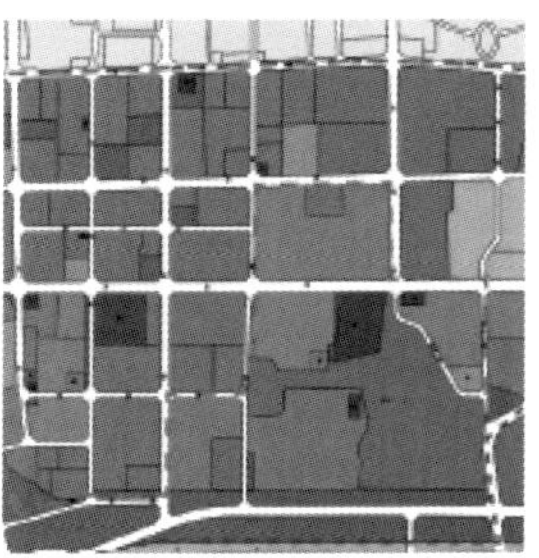

中关村昌平科技园区西区规划调整评估
规划面积：260 公顷
项目位置：北京市昌平区
设计年份：2013 年

中国青年报社综合信息楼
建筑面积：43989 平方米
项目位置：北京市东城区
设计年份：2014 年

石景山区文化中心
建筑面积：41232 平方米
项目位置：北京市石景山区
设计年份：2014 年
合作机构：朱培建筑设计事务所

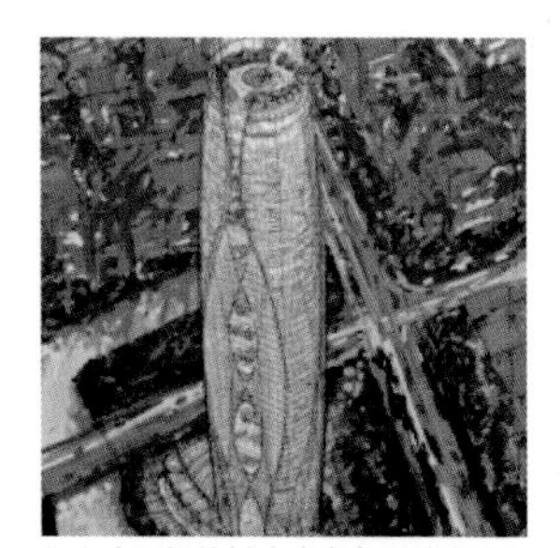

绍兴中国轻纺城中央商务区综合体设计竞赛
建筑面积：162700 平方米
项目位置：浙江省绍兴市
设计年份：2014 年

北京城建北京密码售楼处
建筑面积：1200 平方米
项目位置：北京市大兴区
设计年份：2014 年

陶然亭游泳场及综合楼改扩建
建筑面积：180000 平方米
项目位置：北京市西城区
设计年份：2014 年
合作机构：北京市建筑设计研究院有限公司
艺术中心

北京优联耳鼻喉科医院
建筑面积：69000 平方米
项目位置：北京市朝阳区
设计年份：2014 年
合作机构：香港利安设计集团

吉安市河东滨江新区
控制性详细规划与修建性详细规划方案
规划面积：466 公顷
项目位置：北京市大兴区
设计年份：2014 年

大兴魏善庄（AA-25、AA-29 地块）公建项目
建筑面积：407800 平方米
项目位置：北京市大兴区
设计年份：2014 年

郑州 CSD 配套 E12、E18 地块住宅
建筑面积：293332 平方米（E12）
343036 平方米（E18）
项目位置：河南省郑州市
设计年份：2014 年
合作机构：河南省建筑设计研究院

邢台市龙岗新区概念规划设计
规划面积：2641 公顷
项目位置：河北省邢台市
设计年份：2014 年

中国农机院地块改造工程规划及方案设计
规划面积：7.69 公顷
项目位置：北京市朝阳区
设计年份：2014 年

哈西发展大道周边地区城市设计
用地面积：180 公顷
项目位置：黑龙江省哈尔滨市
设计年份：2015 年

张掖市民政精神病福利院设计
建筑面积：14000 平方米
项目位置：甘肃省张掖市
设计年份：2015 年

北京三里屯创意文荟酒店
建筑面积：21060 平方米
项目位置：北京市朝阳区
设计年份：2015 年

731 部队士官宿舍旧址保护与利用规划设计竞赛
用地面积：19.8 公顷
项目位置：黑龙江省哈尔滨市
设计年份：2015 年

北京丰顺奔驰文化体验中心
建筑面积：8000 平方米
项目位置：北京市丰台区
设计年份：2015 年

麦迪奥雪山酒店
建筑面积：26000 平方米
项目位置：哈萨克斯坦共和国阿拉木图市
设计年份：2015 年

蔓耗古镇概念性规划设计竞赛
用地面积：120 公顷
项目位置：云南省个旧市
设计年份：2015 年

寿县文化艺术中心
建筑面积：19356 平方米
项目位置：安徽省寿县
设计年份：2015 年
合作机构：朱培建筑设计事务所

通州梨园文体中心
建筑面积：300000 平方米
项目位置：北京市通州区
设计年份：2015 年
合作机构：北京市建筑设计研究院有限公司
方案创作工作室

城建怀柔城市设计项目设计竞赛
规划面积：40 公顷
建筑面积：900000 平方米
项目位置：北京市怀柔区
设计年份：2015 年

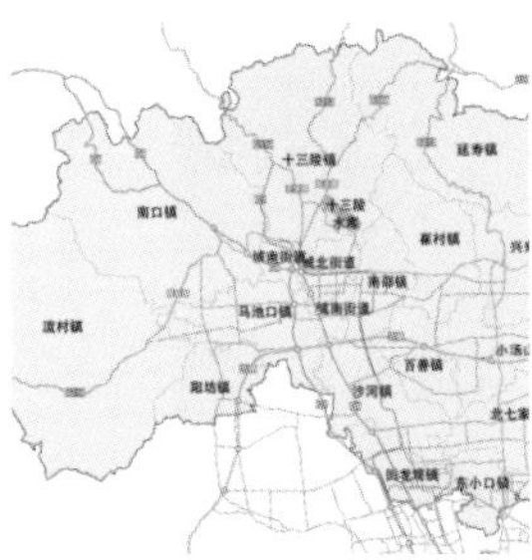

昌平殡葬设施专项规划
项目位置：北京市昌平区
设计年份：2016 年

台湖光机电一体化产业基地 104、109 地块
建筑面积：190000 平方米
项目位置：北京市通州区
设计年份：2015 年

中缅因州医养中心
建筑面积：20000 平方米
项目位置：美国缅因州
设计年份：2015 年

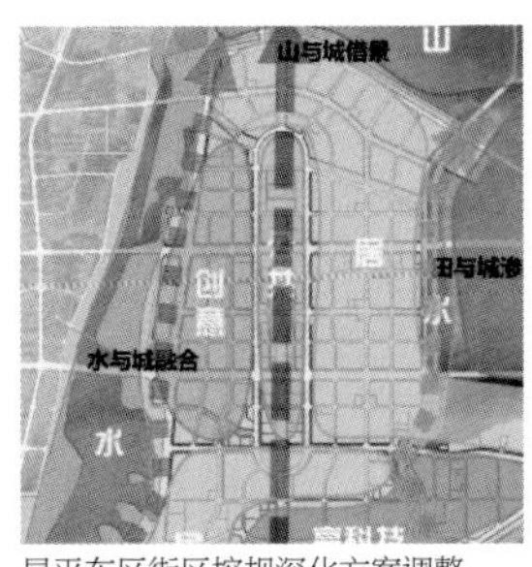

昌平东区街区控规深化方案调整
用地面积：1400 公顷
项目位置：北京市昌平区
设计年份：2016 年

大兴雪花冰箱轨道交通一体化综合开发项目
建筑面积：250000 平方米
项目位置：北京市大兴区
设计年份：2015 年

翠湖湿地博物馆
建筑面积：12000 平方米
项目位置：北京市海淀区
设计年份：2016 年
合作机构：限研吾建筑设计事务所

中关村怀来科技园
规划面积：40 公顷
建筑面积：400000 平方米
项目位置：河北省怀来县
设计年份：2016 年

兰州国际港务区城市设计
规划面积：300 公顷
项目位置：甘肃省兰州市
设计年份：2016 年

远洋密之云项目总体规划
规划面积：15.7 公顷
项目位置：北京市密云区
设计年份：2016 年

白浮泉化庄片区城市设计
用地面积：总体 180 公顷
核心 26 公顷
项目位置：北京市昌平区
设计年份：2017 年

崇礼密苑旅游度假区山地媒体中心（MMC）北区
建筑面积：45000 平方米
项目位置：河北省张家口市
设计年份：2016 年

北京昌平麓鸣花园二期项目
建筑面积：569000 平方米
项目位置：北京市昌平区
设计年份：2016 年

北京・海鹊落原创音乐产业基地概念性规划及城市设计
用地面积：75 公顷
项目位置：北京市昌平区
设计年份：2017 年
合作机构：涌现集团

保千里北京机器人主题园区规划
规划面积：4.6 公顷
项目位置：北京市亦庄经济技术开发区
设计年份：2016 年

唐山国际旅游岛浅水湾改造提升规划设计
规划面积：1200 公顷
项目位置：河北省唐山市
设计年份：2016 年

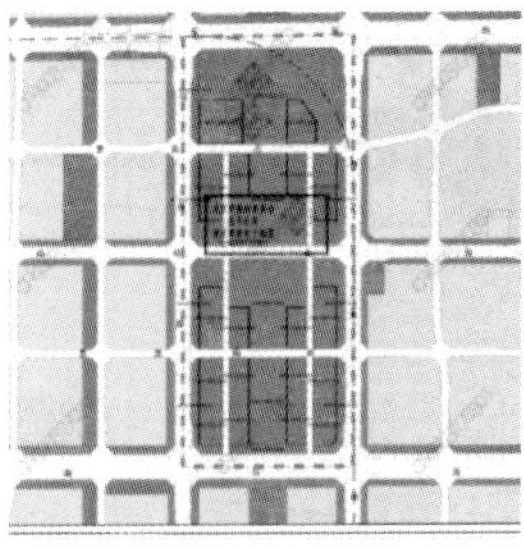
昌平东区中轴区域控规
用地面积：50 公顷
项目位置：北京市昌平区
设计年份：2017 年

橙天嘉禾 360 剧场
建筑面积：9000 平方米
项目位置：北京市朝阳区
设计年份：2017 年

延庆世园会 B 地块住宅
建筑面积：50000 平方米
项目位置：北京市延庆区
设计年份：2017 年
合作机构：北京市建筑设计研究院有限公司第一设计院

西安浐灞自贸中心概念方案设计竞赛
建筑面积：29747 平方米
项目位置：陕西省西安市
设计年份：2017 年

延庆世园会居住组团配套公建
建筑面积：800 平方米
项目位置：北京市延庆区
设计年份：2017 年

延庆世园会数字中心
建筑面积：10000 平方米
项目位置：北京市延庆区
设计年份：2017 年
合作机构：北京市建筑设计研究院有限公司第一设计院

首发东南高速公路智慧物流港规划
建筑面积：300000 平方米
项目位置：北京
设计年份：2017 年

延庆世园会 A 地块住宅
建筑面积：30000 平方米
项目位置：北京市延庆区
设计年份：2017 年
合作机构：北京市建筑设计研究院有限公司第一设计院

延庆世园会安保中心
建筑面积：29747 平方米
项目位置：北京市延庆区
设计年份：2017 年
合作机构：北京市建筑设计研究院有限公司第一设计院

坤鼎南通产业园区
规划面积：8 公顷
建筑面积：100000 平方米
项目位置：江苏南通
设计年份：2017 年

昌平区城南街道化庄社区棚户区改造项目
建筑面积：93000 平方米
项目位置：北京市昌平区
设计年份：2017 年

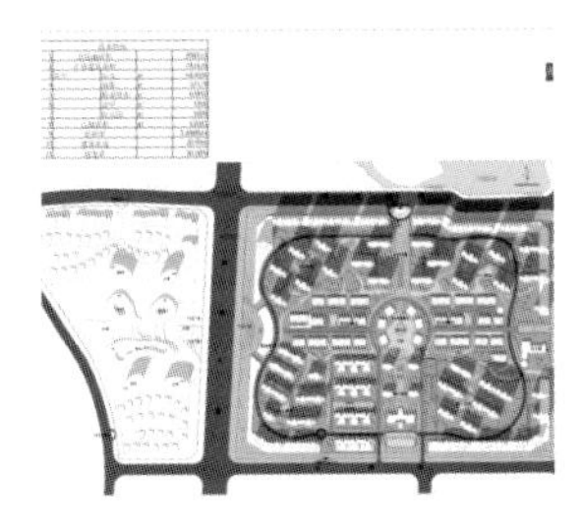

北海中电海湾国际项目
建筑面积：800000 平方米
项目位置：广西省北海市
设计年份：2017 年

北京固安中小学设计竞赛
建筑面积：35654 平方米
项目位置：河北省固安县
设计年份：2017 年

河北阜平照旺台搬迁安置项目
建筑面积：494000 平方米
项目位置：河北省保定市
设计年份：2017 年

昌平区南邵镇 0303-54 地块
F1 住宅混合公建用地项目
建筑面积：375000 平方米
项目位置：北京昌平区
设计年份：2017 年
合作机构：北京市建筑设计研究院有限公司第八设计院

密苑太子滑雪小镇
建筑面积：157088 平方米
项目位置：河北省张家口市崇礼区
设计年份：2017 年
合作机构：新加坡 PAC 城市规划建筑设计有限公司
清华大学建筑设计研究院有限公司简盟工作室

河北阜平四里庄搬迁安置项目
建筑面积：72000 平方米
项目位置：河北省保定市
设计年份：2017 年

崇礼泰禾酒店
建筑面积：30050 平方米
项目位置：河北省张家口市崇礼区
设计年份：2017 年
合作机构：HZS 建筑设计事务所

附录六
参考文献

Appendix VI
Reference

[1] 杜状 . 浅析办公空间中人性化空间的营造 [J]. 艺术科技 ,2017(5).
[2] 刘馨浓 . 互联时代，移动办公 [J]. 建筑知识 ,2015(8).
[3]SMITH，小西 . 空间的力量 法国巴黎 Deskopolitan 联合办公空间 [J]. 室内设计与装修 ,2017(7).
[4] 凯文・林奇 . 城市意象 [M]. 北京：华夏出版社 ,2001.
[5] 范米尔 . 欧洲办公建筑 [M]. 北京：知识产权出版社 ,2005.
[6] 刘文标，刘峰，赵和生 . 创新推动创新——浅谈高新科技园区研发办公建筑设计 [J]. 城市建筑 ,2012(2).
[7]Bill Rigby，Alistair Barr，沈丹琳 . 科技新殿堂：是梦幻总部，还是受到诅咒的标志建筑？[J]. 英语文摘 ,2013（10）.
[8] 田晶 . 后工业社会中的建筑渴望——浅析当代总部办公建筑的设计理念 [J]. 城市建筑 ,2010（8）.
[9]WTL Design. 话题京东亦庄总部 A 栋 [J] . 室内设计与装修 ,2016（8）:52-55.
[10] 芦原义信 . 外部空间设计 [M]. 尹培桐，译 . 北京：中国建筑工业出版社 ,1985.
[11] 老子 . 道德经 [M]. 南京：江苏古籍出版社 ,2001.
[12] 卢向东 . 中国现代剧场的演进——从大舞台到大剧院 [M]. 北京：中国建筑工业出版社 ,2009.
[13] 陈平 . 剧院运营管理——国家大剧院模式构建 [M]. 北京：人民音乐出版社 ,2015.
[14] 庄维敏 . 后评估在中国 [M]. 北京：中国建筑工业出版社 ,2017.
[15] 周怡白 . 中国剧场史 [M]. 北京：中国戏剧出版社 ,2016.
[16] 理查德・谢克纳 . 环境戏剧 [M]. 曹路生，译 . 北京：中国戏剧出版社 ,2001.
[17] 汉斯・蒂斯・雷曼 . 后戏剧剧场 [M]. 李亦男，译 . 北京：北京大学出版社 ,2016.
[18] 杨子涵 . 中国式沉浸——沉浸式戏剧在中国的成长 [J]. 艺苑，2017（1）.
[19] 胡妙胜 . 开拓新空间——环境戏剧的空间设计 [J]. 文艺研究 ,1991（6）.
[20] 胡妙胜 . 戏剧空间结构——舞台设计的美学 [J]. 戏剧艺术 ,1988（4）.
[21] 罗敏杰 . 空间界面视角下的可变剧场建筑设计研究 [D]. 北京：清华大学 ,2013.
[22] 耶日・格洛托夫斯基 . 迈向质朴戏剧 [M]. 尤金尼奥・巴尔巴，编 . 魏时，译 . 刘安义，校 . 北京：中国戏剧出版社 ,1984.
[23] 魏大中，吴亭莉，项端祈，等 . 伸出式舞台剧场设计 [M]. 北京：中国建筑工业出版社 ,1992.
[24] 程翌 . 多维视角下的当代演艺建筑 [M]. 北京：中国建筑工业出版社 ,2015.
[25] 朱小地 . 又见五台山剧场，山西，中国 [J]. 世界建筑 ,2017（3）.
[26] 李道增，傅英杰 . 西方戏剧・剧场史（上）[M]. 北京：清华大学出版社，1999.
[27] 李道增，傅英杰 . 西方戏剧・剧场史（下）[M]. 北京：清华大学出版社，1999.
[28]ATHANASOPULOS G . Contemporary theater：evolution and design，2ed，North Charleston：

BookSurge Publishing ,2006.
[29] 赵国宏 . 为戏剧而存在——戏剧与剧场的空间形式 [J]. 演艺设备与科技 ,2008（6）.
[30] 汉斯・蒂斯・雷曼 . 后戏剧剧场 [M]. 李亦男 , 译 . 北京：北京大学出版社 ,2016.
[31] 中国建筑工业出版社 , 中国建筑学会 . 建筑设计资料集（第三版）第四分册 [M]. 北京：中国建筑工业出版社 ,2017.
[32] 王建国 . 中国城市设计发展和建筑师的专业地位 [J]. 建筑学报 , 2016（7）.
[33] 王建国 . 城市设计 [M]// 中国大百科全书总编辑委员会 . 中国大百科全书 .2 版 . 北京：中国大百科全书出版社 ,2009.
[34] 金磊 . "文化城市"与城市建设的"样本"塑造 [N]. 中国建设报，2017 -05-15.
[35] 傅舒兰 . 杭州风景城市的形成史西湖与城市的形态关系演进过程研究 [M]. 南京 : 东南大学出版社 , 2015.
[36] 王树声 . 结合大尺度自然环境的城市设计方法初探——以西安历代城市设计与终南山的关系为例 [J]. 建筑学报 ,2016（7）.
[37] 缪荃孙 , 刘万源 , 等 . 光绪昌平州志 [M]. 北京：北京古籍出版社 ,1989.
[38]《建筑创作》杂志社，《中国摄影》杂志社 . "风雅运河全国摄影大赛"获奖作品集 [M]. 天津：天津大学出版社 ,2008.
[39] 张剑葳 . 白浮泉、都龙王庙与龙泉寺——京杭大运河通惠河段旧源的建筑与景观 [C]//《营造》第五辑——第五届中国建筑史学国际研讨会会议论文集（下）. 广州：中国建筑学会建筑史学分会 , 华南理工大学建筑学院 ,2010.
[40] 侯仁之 . 古代北京运河的开凿和衰落题 [J]. 北京规划建设 ,2004（1）.
[41] 蔡蕃 . 北京古运河与城市供水研究 [M]. 北京：北京出版社，1987.
[42] 蔡蕃 . 京杭大运河水利工程 [M]. 北京：电子工业出版社 , 2014.
[43] 吴良镛，吴唯佳 . 北京 2049 空间发展战略研究 [M]. 北京 : 清华大学出版社 , 2012.
[44] 单霁翔 . 从功能城市到文化城市 [M]. 天津：天津大学出版社 ,2007.
[45] 宋濂 , 等 .《元史・河渠志》（中华书局点校本）[M] . 北京：中华书局 ,1976.
[46] 宋濂 , 等 .《元史・世祖本记》（中华书局点校本）[M] . 北京：中华书局 ,1976.
[47] 吴仲 . 续修四库全书（史部，政书类）[M]. 上海：上海古籍出版社 , 2013.
[48] 顾炎武 . 昌平山水记 [M]. 北京：北京出版社 ,1962.
[49] 侯仁之 . 北平历史地理 [M]. 北京：外语教学与研究出版社 , 2013.
[50] 侯仁之 . 北京历史地图集 [M]. 北京：文津出版社 , 2013.
[51] 叶楠 , 崔琪 . 大运河北京段遗产保护规划 [J]. 中国名城 ,2011.

[52] 朱颖 . 大运河源考 [J]. 城乡建设 ,2017(14).
[53] 朱颖 . 大运河北段（元）通惠河与白浮“源”的认知 [J]. 建筑创作 ,2017(5).
[54] 凯文・林奇 . 城市意象 [M]. 北京：华夏出版社 ,2001.
[55] 北京市昌平区地方志编纂委员会 . 北京昌平图鉴 [M]. 北京：北京市昌平区地方志编纂委员会 ,2011.
[56]《北京百科全书》总编辑委员会 . 北京百科全书昌平卷 [M]. 北京：奥林匹克出版社 ,2002.
[57] 阳作军 . 趋同与重塑——杭州城市景观的历史演变与规划引领策略 [M]. 北京：中国建筑工业出版社 ,2014.
[58] 孙慧杰 . 从“731”旧址保护规划探讨名城保护新格局 . 黑龙江史志 ,2014（12）.
[59] 海燕 . 美国医院建筑设计新趋势 [J]. 中国医院建筑与装备 ,2015（9）.
[60] 吴培波 . 医院标识导向系统的分类、制作与定位 [J]. 中国医院建筑与装备 ,2015（6）.
[61] 马国馨 . 体育建筑论稿——从亚运到奥运 [M]. 天津：天津大学出版社 ,2007.
[62] 体育产业市场化助推体育建筑精明化营建——马国馨院士访谈 [J]. 城市建筑 ,2015(25).
[63] 周定国 . 农作物秸秆人造板的研究 [J]. 中国工程科学 ,2009,11(10).
[64] 顾道金 . 建筑环境负荷的生命周期评价 [D]. 北京：清华大学 ,2006.
[65]GU LIJING，LIN BORONG，GU DAOJIN，ZHU YINGXIN. An endpoint damage oriented model for life cycle environmental impact assessment of buildings in China[J]. Chinese Science Bulletin，2008，53 (23).
[66] 杨志新 . 北京郊区农田生态系统正负效应价值的综合评价研究 [D]. 北京：中国农业大学 ,2006.
[67] 中华人民共和国国家统计局 .2006 年国家统计年鉴 [M]. 北京：中国统计出版社 ,2006.
[68] 钱杰 . 大都市碳源碳汇研究——以上海市为例 [D]. 上海：华东师范大学 ,2004.
[69] 沈之宇 . 小氮肥氨合成装置先进控制与优化研究与应用 [D]. 合肥：中国科学技术大学 ,2006.
[70] 司景 . 氮肥企业清洁生产评价方法及实例研究 [D]. 合肥：合肥工业大学 ,2006.
[71] 余雷 . 工业企业节能问题研究 [D]. 贵州：贵州大学 ,2005.
[72] 顾道金，朱颖心，谷立静 . 中国建筑环境影响的生命周期评价 [J]. 清华大学学报 (自然版),2006,46(12).
[73] 陈琳 . 农作物秸秆资源综合利用的战略研究——以农作物秸秆人造板产业化发展为例 [D]. 南京：南京林业大学 ,2007.
[74] 张洋 . 麦秸人造板的研究 [D]. 南京：南京林业大学 ,2001.

后记

拾：释义

“拾”是一个汉字词语，作动词时可以指从地上拿起，或拾零、拾取，最常用的是指收、敛、整理，也可以用作数字十的大写。

拾：十年

2007 年 10 月，我和北京市建筑设计研究院原二所所长李海南先生被北京市建筑设计研究院派来参与时名为北京金田建筑设计有限公司（2010 年引入战略投资人后更名为北京建院约翰马丁国际建筑设计有限公司）的重组工作，原建院二所 2A6 工作室成员和两名财务人员大约 10 人共同组成了今天北京建院约翰马丁国际建筑设计有限公司的班底，截至今年正好十年了。

拾：整理

团队一直在探索和思索中前行，自成立以来，对公司的环境观和城市观的探索、文化引领设计手法的研究、新建筑技术的研发、团队设计观的总结一直都在进行中，这些都为公司创造更多精品提供了理论支持，公司对于文化引领设计的思索，也使公司提出的“运河源”观点在后申遗时代的大运河文化挖掘中有极其重要的现实意义。

拾：十记

在 2014 年公司推出了公司的第一本品牌手册之后，本次我们整理出更能代表我们设计思想的十篇记录件文章、随笔，结集而成《建筑拾记》，虽有敝帚自珍之嫌，但也希望可用“拾”引发更多思考，是为十记。

致敬

《建筑十书》是建筑学的经典著作，于公元前 27 年由古罗马建筑师维特鲁威著，千百年来，一直是建筑学教育的必读书目之一。《建筑拾记》同样是一个积极求索的设计团队的思索集，涉及的内容不仅仅是狭义的建筑设计，也包括广义的建筑设计，如城市规划、建筑历史、绿色建筑、室内设计、建筑技术、建筑安全、建筑材料等，还涉及很多交叉学科，包括美术、音乐、戏剧、水利、心理学、运营管理等相对宽泛的领域，在 2000 年后的今天，团队谨以《建筑拾记》向维特鲁威致敬。

致谢

向团队中每一位执着的建筑师、工程师致谢，向每一位曾经为公司的成长付出艰辛和汗水的同事致谢，向团队中继续砥砺前行的每一位致谢，向团队背后给予宽容和理解的每一位同事的家人致谢，向每一位给予团队关爱和支持的师长、同人、朋友致谢，更要向支持信任我们并给予我们厚爱的业主致谢。

感谢北京市建筑设计院有限公司的领导、技术专家、同事和曾给予我们帮助的各个合作团队，我们的每一个小的进步都离不开北京建院的支持。

感谢美国约翰马丁工程设计集团总裁威勒·马丁先生和罗超英对于整个管理团队的信任和支持。

感谢领我踏入建筑之门的吴观张、王昌宁先生，感谢全国勘察设计大师何玉如、吴亭莉伉俪，刘力、周文遥伉俪以及张宇、邵韦平等建筑大师多年来给予我们建筑创作上的引导。感谢著名医学专家中国工程院院士韩德民先生、著名音乐人李鹿女士给团队的指导。

感谢著名建筑师、艺术家朱小地先生多年来给予团队的指导与关爱。他带领我们完成的又见五台山项目使团队受益匪浅，他一直是团队的榜样。

我还要特别向在给予我们指引的同时，于百忙之中提笔为本书作序的北京建院党委书记、董事长徐全胜，中国工程院院士马国馨先生表示诚挚的谢意，所有这些都是对我们这个团队最大的鼓舞。

最后要向参与本书策划、编辑的《中国建筑文化遗产》编辑部表达谢意，金磊主编、殷力欣副主编、李沉副主编、苗淼助理、朱有恒主任、董晨曦副主任等均为本书的出版做出了贡献。同时，感谢天津大学出版社原副社长韩振平、郭颖编辑的支持。还要对建院马丁公司中为了本书的出版默默奉献的同事表示感谢，诚然没有大家的共同努力本书是难以问世的。

朱颖
北京建院约翰马丁国际建筑设计有限公司
董事长、总经理
2017 年 12 月 30 日

Afterword

Shi: definition

Shi is a Chinese character, which means picking up something from the ground or collecting things when used as a verb. It often means collecting or sorting. When used as a noun, it means ten.

Shi: ten years

In October 2007, Li Hainan, the former director of the 2nd Design Department of Beijing Institute of Architectural Design (BIAD), and I were sent by BIAD to participate in the reorganization of the Beijing Jintian Architectural Design Co., Ltd. (renamed as BIAD John Martin International Architectural Design Co., Ltd. (the Company) after the introduction of the strategic investors in 2010). Back then, approximately ten people including the former members of the 2A6 Studio in the 2nd Design Department of BIAD and two financial staff constituted the original team of the Company and embarked on a new journey. It has been exactly ten years since the reorganization.

Shi: sorting

We have always been feeling our way as we go. Since the very beginning, we have made explorations to forge our unique corporate views about environment and cities, studied how to base our design techniques on culture, developed new architectural technologies, and summarized what our team achieved into design concepts. All of these efforts provide theoretical support for the Company to design more masterpiece works. Thanks to the Company's thinking on culture-based design, the view of "canal source" proposed by the Company is of extremely important practical significance to the exploration of the Grand Canal culture in the age of post-declaration on world cultural heritage.

Shi: ten gathered notes

After releasing our first brand brochure in 2014, we publish the *Shi Notes on Architecture* this time by sorting out ten essays, research records, articles on scientific and technological innovation and casual notes, all of which epitomize our design ideas. This book is far from being complete and only for your reference.

Homage

The *Ten Books of Architecture* is one of the classic works for architecture professionals written by the Roman architect Vitruvius in 27 BC. This book has been a must read for architecture education for thousands of years. The *Shi Notes on Architecture* is a treatise on architecture written by our aspiring design team. It covers architectural design in both the narrow and broad senses, such as urban planning, architectural history, green building, interior design, building technology, building safety and building materials, as well as a wide range of interdisciplinary areas such as fine arts, music, drama, water conservancy, psychology, and operation management. We pay homage to

Vitruvius 2000 years later by this book.

Acknowledgement

I would like to extend my gratitude to the dedicated architects and engineers in our team, to our colleagues for their hard work underpinning the growth of the Company, to the team members for their persistent progress, to the families behind our team for their tolerance and understanding, to the teachers, peers and friends for their care and support, and to the owners for their trust and concern.

I would like to extend my gratitude to BIAD's leaders, technical experts, colleagues and cooperating teams that have helped us, for their support in every achievement we have made.

I would like to extend my gratitude to the Chairman of John A. Martin & Associates, Inc. (JAMA) and Ms. Luo Chaoying, Vice Chairman of JAMA for their trust in our management team.

I would like to extend my gratitude to Wu Guanzhang and Wang Changning who lead me to this industry, to our survey and design masters He Yuru and his wife Wu Tingli, to Liu Li and his wife Zhou Wenyao, as well as to master architects Zhang Yu and Shao Weiping, for their years of guidance on our architectural design. Special thanks to famous medical expert Academician Han Demin and Famous musician Ms. Li Lu for their guiding.

I would like to extend my gratitude to Mr. Zhu Xiaodi, a famous architect and artist, for his guide and care to our team over years. Our design team benefited from the work he led us to do. and his dedication has always been a role model for us.

I would also like to express my sincere appreciation to Mr. Xu Quansheng, the Party Committee Secretary and Chairman of BIAD, and Mr. Ma Guoxin, the Academician of the Chinese Academy of Engineering, for guiding our work and taking the time to write prefaces for this book. Their helps are the biggest encouragement for our team.

In the end, I would like to extend my gratitude to the Editorial Department of China Architectural Heritage who participated in the planning and editing of this book. The Editor–in–Chief Jin Lei, the Associate Editor–in–Chief Yin Lixin, the Associate Editor–in–Chief Li Chen, the Assistant Miao Miao, the Director Zhu Youheng, and the Deputy Director Dong Chenxi, have made contributions to the publication of this book. I would also like to express my appreciation to the former Vice President Han Zhenping and the Editor Guo Ying of Tianjin University Press for their support, and to our colleagues for their contribution to the publication of this book. This book comes as a result of the joint efforts of all the people mentioned above.

Zhu Ying
Chairman and General Manager of BIAD
John Martin International Architectural Design Co., Ltd.
December 30, 2017

《建筑拾记》编委会